Marc van der Erve is a writer whose career as advisor spans several decades. He holds a BSc in *Applied Physics* and a PhD in *Sociology*. His academic focus ranges from non-linear systems to the spontaneous rise of forms of human organization. With a keen eye on the needs of societal leaders, he explores both the practical and the philosophical on the crossroads where the natural and social sciences meet.

MARC VAN DER ERVE

THE NEXT SCIENTIFIC REVOLUTION

AND HOW IT MIGHT SHAPE THE 3RD MILLENNIUM

The Next Scientific Revolution
Antwerp/Somerset West, 2013

Copyright © Marc van der Erve, 2012-2013

For more about the author and his work, visit:
http://www.marcvandererve.com

ISBN Paperback: 978-0-620-50969-5
ISBN Digital Version: 978-0-620-50970-1

"Effective scientific research scarcely begins before acquiring answers to questions like: What are the fundamental entities of which the universe is composed? How do these interact with each other and with the senses?"
Thomas Kuhn

Foreword

Early in life, I developed a hunch, against the odds, so it seemed, that all we see in our world can be reconciled without the need of getting down to levels where particles and physicists rule. The "Holy Grail", so to speak, is right there in front of our eyes if we only know how to distinguish it. I forgot about it until much later when my study of evolving organizations led me to questions that hadn't been answered. In pursuit of an explanation, I broadened my research and explored various scientific fields. At some point, when evaluating the path that I had followed, I was not so surprised to discover that scientists and philosophers had identified bits and pieces of this "Holy Grail" enough, in fact, to unveil the remainder. This book is autobiographical in the sense that it tells the story of my journey, my setbacks and eventual triumph in uncovering a remarkable perspective of our world.

Marc van der Erve

Contents

Introduction

There are numerous ways to view our world but, on the whole, we view our world as a "material" phenomenon. To explain our experience of thought, many believe that our material world is accompanied by a spiritual world. For thousands of years, this tandem has determined what we think and do. But what if we are wrong? What if our world is not what we think it is?

There are good reasons to believe that this is so. Faint but fundamental findings have popped up in the world of science, findings that are like dots in a puzzle. Connect these dots in a certain way and a novel picture of reality appears, a picture so familiar, yet so different that it will deeply affect the way we interpret our world, if true. Because "reality" is involved, this picture could give new impetus to scientific research as much as it could inspire leaders to rethink their role.

This book is not an essay of philosophy but a practical analysis of developments across the natural and social sciences. It explains how the established system of ideas at the heart of the current worldview may be making way for a new system, a new "paradigm", if you like, that will dictate how we see, study, and manage our world in future.

In the first part of this book, I'll explore why we often do not perceive something even when looking it in the face. I will conclude with a glimpse of this new worldview. In the second part, I'll explain how I stumbled onto dots that led me to identify the outlines of a new picture of reality. In the third part, I'll explain the crux of this new picture and why it might set off a scientific revolution. I'll also discuss how it matches the ideas of established thinkers, refraining from a wider analysis because this is beyond the purpose of my book. In the fourth and last part, I'll list the main trends that will ripple through society should the new worldview find traction.

Of course, a paradigm shift remains speculation until it materializes. Nonetheless, this book is relevant to leaders of science and society because it helps them imagine the broad societal drifts that may be ahead of them. This book is also useful to undergraduate and post-graduate students because it puts into perspective "normal science", that is, the science that they are being made familiar with.

Do You Get What You See?

When someone was kind enough to give me a copy of the biography of Steve Jobs, one of the two founders of Apple who had died of pancreatic cancer months before, I was keen to read it. [1] Since 2004, I had followed Apple after publishing an article in which I analyzed the fate of several companies and, as it turned out later, with some success. [2] Having finished reading the book, I asked myself what the "essence" of Jobs had been. Sure, he was what I called in my article a "builder-type leader" who ended up transforming the personal-computer industry. But, what really explained his "being"? Three things came to mind. First, Jobs rose to the surface in an exceptionally fertile environment: the 40-mile Santa Clara Valley between San Francisco and Palo Alto, which was labeled "Silicon Valley" by columnist Don Hoefler in 1971. "The Valley" had fostered formidable companies, such as Hewlett Packard and Intel. As the author reported, Jobs was inspired by "the history of the place" and "wanted to be part of it". Second, the crux behind Jobs' success was his distinctive and consistent behavior. Obsessed with perfection, Jobs challenged people that worked for him, driving them to deliver far beyond what they believed possible initially. Third, Jobs had an exceptional sense of what customers wanted. In his own words, "Customers don't know what they want until we've shown them." Thus, Jobs early perceived and was driven by "market potential".

Near the end of 2011, three months after the death of Jobs, the community of scientists working at the "Large Hadron Collider" in Geneva reported that they had seen glimpses of the Higgs boson, a much-anticipated elementary particle predicted by the Standard Model of particle physics. Envisaged by the British physicist, Peter Higgs, and other scientists, Higgs bosons are believed to swarm massless particles that speed through the universe. When they interact with these particles, they hinder their movement. As a result, the movement-related (or kinetic) energy of these particles is converted into what scientists call "mass". So, Higgs bosons help nature's tiny building blocks gain mass. The Higgs boson is popularly referred to as the "God particle" not just because of its crucial role but particularly also because it is so difficult to observe. It decays into other particles before the massive detectors of the Large Hadron Collider can record it. The traces of these particles can be seen, however. So, in the end, *they* confirm the existence of the Higgs boson.

The story of Jobs must have been lingering in my mind when a parallel between the existence of Jobs and the Higgs boson dawned on me. First, once scientists affirm the existence of the Higgs boson, it will have risen to the surface in an exceptionally fertile environment: the Large Hadron Collider, a multi-billion dollar project that involved boring a circular 27-

kilometer tunnel 175 meters underground. The Large Hadron Collider is capable of accelerating opposing particle beams close to the speed of light. When these highly energized beams collide, particles break down into a multitude of other particles at impact. The collider's huge detectors record the traces of these particles including the traces of particles that emerge through spontaneous decay. Second, once affirmed, the Higgs boson will have shown itself through repeated and statistically consistent decay in line with scientist predictions. At the end of 2011, the recordings of some 350 trillion collisions had been analyzed with only ten collisions producing evidence of the Higgs boson.[3] The evidence, however, was found in two of the collider's main detectors, each located in a different part of the tunnel. Third, in both detectors, evidence of the Higgs boson was expected to appear (and now has appeared as it seems)[4] at a distinct energy level: an energy potential difference of some 125 billion electron Volts, the latter representing the "units" that scientists use (instead of "kilos") to measure the mass of elementary particles, such as the Higgs boson.[5] In all, the essence of the reality of both Jobs and the Higgs boson seems to hinge on: local environment, consistent behavior, and some potential difference, such as a market or energy niche.

All of this sounds philosophical, of course. I guess some readers will be inclined to remind me that the reality of Jobs or the Higgs boson essentially is a matter of matter. But, is it? The mathematical description of the Higgs boson and its interaction with other particles revolves around waves, spin, and motion. So, the Higgs boson and, for that matter, other particles appear to be behavioral phenomena rather than "things". What is more, the conduct of Jobs followed a complex of such "behavioral phenomena" (or particles) that came together through a much bigger "project", which started a little less than 14 billion years

15

ago.[6] It involved the emergence of "everything", from nature's tiny building blocks to stellar systems, life, human society and "The Valley". In the end, it produced traces that were Jobs'. Nonetheless, when, in a dialog, I referred at some point to particles as non-material behavioral phenomena involving waves, spin, and motion, someone simply asked: "The waves, spin and motion of *what*?"[7]

Inhabitant Or Prisoner?

The common assumption is that our world is ultimately made up of points in space or *things*.[8] What else would "wave, spin or move"? So, when we are confronted with Werner Heisenberg's much-quoted *uncertainty principle*, which suggests that particles behave as either waves or things, most of us tend to believe the balance will tip to the latter. This is not surprising. About 2,300 years ago, Aristotle, the founder of Western philosophy, taught that our world involved "substances, which are divisible into other substances."[9] In the 17th century, the French philosopher, René Descartes, depicted a similar view in his book *Treatise on the World*. According to Descartes, the universe is made up of tiny "corpuscles of matter" that can be divided (and assembled) into other corpuscles.[10] This view permeated the assumptions of subsequent scientists. Isaac Newton, for example, believed that light consisted of small particles (rather than waves), a view that dominated the opposite findings of the Dutch scientist, Christiaan Huygens, a contemporary of Descartes.[11] Huygens' discovery of the wave-like nature of light was finally confirmed two centuries later.[12] In the light of the latter, Albert Einstein made an intriguing contribution for which he received the Nobel Prize in 1921. Einstein suggested that light waves consist of discrete "quanta" of energy (later called photons). At the same time, he showed that the energy of each quantum is related to the frequency of the light wave. This is intriguing for two reasons. First, Einstein confirmed that light consists of waves. However, by proving that light waves come in distinct quanta of energy, he brought in the notion of the particle again.[13] So, in spite of the growing evidence of the wave-like nature of light (and other particles), the idea of a world of things had all but disappeared. In 1967, the journalist and social philosopher, Arthur Koestler, introduced a view, which further stimulated the idea that our world is a world of things.[14] Koestler expanded Aristotle's idea of "substances that are divisible into other substances". Inspired by the structure of complex systems, such as living organisms and human organizations, Koestler coined the term "holon" to refer to something that is both *a whole* and *a part*. For example, a cell is a stable phenomenon that is both *a whole* and *a part* of some organ. A business unit or department is both *a whole* and *a part* of some company. In other words, Koestler invented a universal label for things in the hierarchy that makes up our world. Although the invention of the "holon" is evidence of the human need for a more integrative insight into what we experience as reality, it did not essentially change the common view of our world as a world of things (that can be divided and assembled). It seems that, no matter what our field is, we tend to look at our world through a common fence, a grid of assumptions that, in view of the earlier-hinted reality of Jobs and the Higgs boson, may well prevent us from truly "grasping" it. As a result, we may not just be inhabitants of the world that we observe, but practically also its prisoners.

Jumping Fences

As common bodies of assumptions that determine how we view our world, fences are by no means rare. The American historian and philosopher of science, Thomas Kuhn, investigated various examples of such fences, referring to them as "paradigms". Kuhn defined a paradigm as "a recognized universal body of scientific achievements that, for a time, provides model problems and solutions to a community of practitioners".[15] One of Kuhn's most prominent "fence" examples involved a model of the universe that ruled the thinking for at least fourteen hundred years. This model was at the heart of a system used to calculate the position of planets and stars. Claudius Ptolemy, a Roman scholar who lived in Egypt, formalized this system in the 2nd century. At the heart of Ptolemy's system was Aristotle's *geocentric* model of the universe, which had the Earth at its center. From Apollonius of Perga, another Greek thinker, Ptolemy adopted the idea that each planet moved in a small circle, which itself moved along a larger circle around the Earth as did the Sun.[16] Ptolemy needed these "circles within circles" to bring his system in line with actual observations and improve its accuracy. Ptolemy further calibrated the system by placing the Earth slightly off center. The remarkable success of Ptolemy's system as a tool for planetary observation explains why the model of the universe behind it ruled the thinking of leaders of science and society until the 16th century when the Polish astronomer, Nicolaus Copernicus, offered a feasible alternative. As such, the *geometric* model of the universe proved to be a grand historical fence that prevented leaders from "grasping" their world differently.

So, what led to the proposal of a different model of the universe, one that had the Sun rather than the Earth at its center? In other words, what led to jumping Ptolemy's fence? Kuhn pointed out that, as astronomical observations became more accurate, "Ptolemy's system never quite conformed with the best available observations". Successive generations of astronomers were able to deal with such discrepancies by adding more circles to the system, making it more complex. In the end, the system's "complexity was increasing far more rapidly than its accuracy." Kuhn concluded that this convinced astronomers early in the 16th century "that the astronomical paradigm was failing." As it turned out, the Renaissance produced a scientist, Copernicus, who formulated a complete *heliocentric* cosmology in which the Sun displaced the Earth from the center of the universe. Copernicus argued that a universe with the Sun at its center would make matters simpler. In Ptolemy's system, for example, each planet was treated differently. Copernicus arrived at a more unified approach, which also yielded the correct arrangement of planets starting with the planet, Mercury, closest to the Sun. [17] However, Copernicus continued using the circles-within-circles system of Ptolemy and failed to produce more accurate predictions. As a result, rather than simplifying the system, he ended up making it more complex. This explains why his contemporaries rejected the model. The paradigm shift or, as Kuhn also called it, the "scientific revolution" that Copernicus' model set in motion took much longer to complete. Sixty-six years later, the German astronomer, Johannes Kepler, identified a system of ellipses that accurately described the planetary trajectories in Copernicus' model. Final resolution came seventy-eight years after that when Newton introduced the law of universal gravitation and proved the consistency of Kepler's laws of planetary motion.

All in all, Kuhn's analysis of the "Copernican Revolution" suggests that anomalies point to a road, which leads to a better model if, indeed, such a model does exist.[18] So, considering my earlier speculation about the behavioral origin of reality and assuming the viability of this idea, the way to start is to identify anomalies, dots that, when connected, might reveal reality's true shape (something that I undertake in the next part of this book). Then again, as long as the belief persists that these anomalies can be explained within the boundaries of established thinking, the "world of things" will keep us in its spell. What is more, research showed that anomalies and different perspectives (this is what models and paradigms basically are) are often not even rejected but simply not observed. Fortunately, indirectly, this finding also produced a *catalyst of recognition*.

Fence-Jumping Psychology

Not just the acceptance of a new scientific model but also the acknowledgement of anomalies that hint at it is problematic. Anomalies are often not perceived as anomalies. Kuhn referred to the research of the American psychologists, Jerome Bruner and Leo Postman, who investigated this phenomenon.[19] In an experiment, Bruner and Postman asked students individually to identify a selection of four playing cards, exposing only one card at a time. Whereas most cards were normal, some had been tweaked by reversing their color (e.g. a black three of hearts, a red two of spades). The experiment showed that normal cards were usually identified as normal. But, remarkably, so were the trick cards. After increasing the exposure time, the students became aware that some cards were not what they were supposed to be. Only after increasing the exposure time further, would most students, often suddenly, identify the trick cards, not to be tricked again. Yet, some students never reached that point no matter how long the exposure time. As Bruner and Postman stated, this "reaffirmed that perceptual organization is powerfully determined by expectations built upon past commerce". In practical terms, this means that when we don't expect to see something (not having learned about it before), we generally don't "see" it. The relevance of the findings of Bruner and Postman to the discussion here is not just in the assertion that anomalies are often not perceived as anomalies. Indirectly, they also suggest that an exploration of "what to expect" is vital both to the recognition of anomalies as anomalies and to the recognition of the paradigm that these anomalies may hint at. So, the outlines of a new paradigm of reality further on and in Part 3 (one with the behavioral origin of reality at its center), the discussion of anomalies or dots in Part 2, and the discussion of the resulting societal trends in Part 4 should be considered as projections of "what to expect". As catalysts of recognition, these projections are crucial to leaders of science and society.

What is a fence-jumping or paradigm-shifting experience actually like? Kuhn points to our capacity to recognize visual shapes or forms, which involves a mental process that, like the acknowledgement of anomalies, is shaped by experience. The domain of Gestalt psychologists, visual recognition is about "the essence or shape of a thing" or, in a word, "Gestalt". For instance, the essence or shape of a thing in a picture (its Gestalt) depends on how a person interprets the lines and blotches that make it up. A well-known example involves a picture in which some see a young lady looking away and others an old lady looking down. Another example is a picture that might either be interpreted as the head of a duck or as the head of a rabbit. Generally, we only see *one* essence or shape until someone points out the other possibility. From then on, we can switch back and forth between the two possibilities (or essences) at will. In the light of this discussion, Arthur Koestler provided an apt observation related to the model of the universe. As he noted, "Copernicus knew no more about the motion of planetary bodies than Ptolemy had known".[20] So, both Ptolemy and Copernicus looked at the same picture. Yet, the essence of what they saw was fundamentally different. Ultimately, to paraphrase Kuhn, a fence-jumping or paradigm-shifting experience means "picking up a stick by the other end, which involves handling the same bundle of data as before but placing it in a new system of relations".

Figure 1 The Rubin vase with *figure* (left) and *ground* (right).

Reality appears to involve a duality of Gestalts too. It either concerns *things* or, as the examples of Jobs and the Higgs boson show, it is a *behavioral phenomenon that produces our perception of things*. The "Rubin vase", named after its inventor, the Danish psychologist, Edgar Rubin, is an example that I found particularly useful when explaining this duality. The vase, left in Figure 1, represents reality, as we currently perceive it, that is, a reality of *things*. On the right, you'll see the outlines of two *symmetric* faces that shape the vase. These faces represent the Gestalt of reality as a *behavioral phenomenon that produces our perception of things*, such as the vase. Quite fittingly, Rubin refers to the vase (on the left) as "figure" and to the two faces (on the right) as "ground". The duality of reality Gestalts also comprises *figure* and *ground*, the *figure* Gestalt representing the *symptom* and the *ground* Gestalt representing the *cause*. Indeed, as I hope to show the reader in this book, the perspective of reality as a *behavioral phenomenon that produces our perception of things* truly *is* the ground of reality.

Cracks In Today's Paradigm

If asked today to describe the reality of a snowflake, we would intuitively focus on the *figure* Gestalt of reality, that is, the reality of *things*. We would rely on the ancient guidance from Aristotle, who stated, "substance is plainly divisible into other substances. If not substances then there must be attributes."[21] Of course, the *substance* of a snowflake involves frozen water molecules, that is, assuming that pollution did not leave foreign substances in the air where the snowflake emerged. We further know of water molecules that they are divisible into one large oxygen atom and two smaller hydrogen atoms. It does not make sense dividing these substances any further because this would not add to the description of the snowflake's reality. Aristotle now suggests looking at the *attributes*. For one, a snowflake has a crystalline structure, which grows from water vapor below 0° Celsius and is made up of hexagonal lattices of water molecules. However, due to the variety of atmospheric circumstances, no snowflake that falls from the sky is the same. The Japanese physicist, Ukichiro Nakaya, discovered a potentially meaningful relation between the temperature and the shape of a snowflake. In due course, the American physicist, Ken Libbrecht, confirmed Nakaya's findings. Libbrecht grew snowflakes in a controlled environment[22] and showed that, at a particular temperature, snowflakes consistently grow in a distinct shape. On his website, Libbrecht admits that he doesn't really know why this is so.[23] He evaluated in detail attributes that might play a role during the formation of a snowflake, such as particle diffusion, surface attachment, capillarity, heat diffusion, and the process by which the facets of hexagonal ice crystals develop.[24] The legitimate premise of thinking today is that these rather mechanistic attributes help explain the assembly process of water molecules when the transition from vapor to solid takes place. Not having quite succeeded, Libbrecht noted that the complexity of the physics involved had resulted in a rather large gap between crystal-growth theory and the practice of growing ice crystals. "A great deal remains unexplained even at a qualitative level." Powerfully focused on the *figure* Gestalt of reality, Libbrecht re-affirms, "I would like to understand the fundamental physics of how molecules jostle into place to form a crystal. How does it change with temperature? What happens if there are chemical impurities on the ice surface?"[25] The shape-inspired (or *figure* Gestalt-based) assembly of water molecules is intensely studied these days.[26] However, as Libbrecht observes, the assembly of snowflakes (how they form) is still far from solved. "We have to marry the mathematics with the physics, and that has not been done, partly because we don't know the right physics." [27]

The crux of the problem with today's paradigm of reality is the *assembly process*. As Libbrecht showed, it is exceptionally difficult to describe how floating water molecules self-assemble into *consistent* shapes by chance without tiny workers (or, as Aristotle hinted, an unmoved mover or God) at the assembly line making choices based on the attributes involved. Of course, the idea of nature "constructing" things by itself means we are mistakenly projecting our own capabilities on our world. The invention of a "constructal law" serves as further evidence of such inclination. [28] It explains, for instance, that the shape of a river reflects its path of least resistance. Not even wrong, such a law seems silly in that it still rests on a mechanical premise and fails to answer the question of assembly. In all, Libbrecht made a most important contribution. Without realizing it, he not only identified cracks in today's paradigm of reality but also pointed the way to its *ground* Gestalt.

A New Paradigm Of Reality

The *ground* Gestalt of reality is the new paradigm of reality. The structure of a snowflake is like Rubin's vase. It gets its shape from observable traces of some behavioral phenomenon like the realities of Jobs and the Higgs boson did. In other words, the *essence* of reality is its *ground* Gestalt. Without it, no shape (or thing) emerges. As indicated before, it revolves around: *local environment*, *consistent behavior*, and some *potential difference* or *energy niche*. Conveniently, Libbrecht, directly and indirectly, identified all of these. The *local environment* involved the air saturated to a specified level with *pure* water droplets. The *energy niche* or *potential difference* is the temperature of the vapor cloud, which reflects the energy of the water molecules in the cloud. What remains is *consistent behavior*. Once we know that, we know what shapes a snowflake.

The energy of water molecules involves motion.[29] When a water molecule from the cloud settles on an emerging snow crystal, it freezes and, thus, no longer moves freely in the air. Caught in a lattice, it doesn't rotate either. It only vibrates. Its vibration behavior is determined both by its energy and by its constituents. The union of one large oxygen atom and two small hydrogen atoms produces a wobbly vibration pattern. The 1977 Nobel laureate, Ilya Prigogine, showed that nature favors a *state of least resistance*.[30] Accordingly, as explored further in Part 3, a crystal shape develops when vibrating water molecules engage in a least-energy fitting dance. In this way, snowflakes emerge from the least-energy fitting dance of millions of vibrating frozen water molecules not unlike the complex shapes that emerge through a straightforward add-on process involving a simple figure in the computations by the British physicist, Stephan Wolfram.[31] So, the temperature of the cloud determines the vibration pattern of frozen water molecules and, indirectly, their consistent, least-energy fitting behavior. The latter leaves a trace in the form of a snowflake. In sum, consistent behavior produces consistent shape. Or, more accurately, the prevalence of least-energy fitting behavior patterns produces symmetry.

The *ground* Gestalt of the snowflake's reality involves least-energy fitting molecule vibration patterns. In the end, *they* determine the snowflake's shape. So, an analysis of the effect of environmental conditions should start with how these influence the vibration patterns of the molecules involved. Computations involving some least-energy fitting procedure will then show how certain vibration patterns, including those of impurities, affect the snowflake's shape and, thus, the *figure* Gestalt of its reality. As the examples further on in this book illustrate, when reaching for the *ground* Gestalt of reality, we reduce our world to a sentient harmony of behavior patterns, of which the rise and decline can be conducted and, possibly even, orchestrated. Rather than an assembly process, it involves behavior patterns that, driven by some potential difference, energy niche or inequality, engage in a way that favors the least-energy fitting.[32] Reducing our world this way does not narrow our perspective but, rather, broadens it to a range of behavioral commodities that helps us grasp more uniformly the evolving variety of nature. As Kuhn observed, "in science, reduction is desirable because it explains why and how something exists". The *ground* Gestalt of reality thus allows us to jump the fence.

If Not A Thing Then What?

According to Kuhn, effective scientific research scarcely begins before identifying the *basic entities* of which our universe is composed. Within the current framework of thinking, this basic entity is a thing (e.g. a vase) or, ultimately, as physicists call it, a point in space. In fact, the *figure* Gestalt of reality makes us see our world essentially as a material phenomenon, a hierarchy of things (that can be both observed and assembled), of which the growing branches become more diversified in time. Chance as well as deliberate action brings things and attributes together into novel assemblies through a variety of processes (see Table 1). In contrast, the *ground* Gestalt of reality does not hinge on things but on behavior patterns. Of course, it is difficult to imagine a behavior pattern that might serve as a basic entity without involving a thing that moves. It is possible nonetheless. After all, rather than just behavior, the *ground* Gestalt of reality involves *local environment*, *consistent behavior* and *potential difference*. The interaction between these elements helps us identify the basic entity of what essentially is a non-material phenomenon. When a certain *potential difference* or *inequality* emerges in an environment, it invites orderly, least-energy fitting behavior patterns that try to minimize the inequality as fast as the local conditions allow. [33] For example, if water molecules settle on a snow crystal at five degrees below the freezing point of water then the latter represents a temperature inequality. The water molecules that settle will join the least-energy fitting dance possible to free up energy and minimize this inequality. So, when an inequality develops, it triggers behavior patterns that try to minimize it. With this in mind, the basic entity we seek must be the closest possible relation between inequality and behavior pattern. This is the *inequality-border dance* itself. An inequality border resembles a flag that flutters in a breeze. A flag flutters or dances as a result of the changing inequalities between the flows on either side. Imagine this fluttering without the flag and you get what looks like an "inequality-border dance". At the level where string theory explains how elementary particles emerge from the consistent behavior patterns of strings, we'll probably find inequality-border dances that might qualify as basic entity. The string itself seems like an obvious candidate. The behavior pattern of a string indeed resembles the vibration pattern of frozen water molecules. The latter shapes a snowflake, the former an elementary particle. Then again, as I'll discuss further on in this book, strings, as they have currently been defined, probably do not qualify. [34] In sum, the *ground* Gestalt of reality shows how the universe of things that we observe emerges from a unified, non-material phenomenon. The latter involves inequality-border dances and a grand hierarchy of such dances or behavior patterns that can be detected generally only when these patterns reproduce. The *ground* Gestalt of reality depicts nature as an eternal process of becoming that meanders between rise and decline on a path of least resistance (see Table 1).

	Paradigms of Reality	
	Figure Gestalt	*Ground* Gestalt
Essence	Material	Non-material
Basic entity	Point in space	Inequality-border *dance*
Hierarchy of	Things, shapes	Behavior patterns
Observable	Practically always	Only when reproducing
Existence	Relatively stable	Emergent
Established by	Assembly	Least-energy fitting
Cause	Creation or chance	Inequality minimization
View of nature	Diversified	Unified
Nature	Involves processes	Is process

Table 1 The old (*figure*) and new (*ground*) paradigm of reality.

Kuhn noted about paradigm shifts, such as the shift from the *figure* to *ground* Gestalt of reality, that successive paradigms are often not "commensurable", which means they are not measurable or comparable by a common standard. As a result, a new paradigm cannot be proven by the rules that govern in the old paradigm. On the other hand, a paradigm does not have to conflict with its predecessor. It might deal with phenomena not previously realized and may guide research even in the absence of rules. As the Copernican revolution showed, it may indeed take years before such rules are identified.

Empathy Is All You Need

Nature doesn't think in terms of "inequality minimization" or thermodynamic laws. It doesn't have a clue of our abstractions. As explained further on in this book, it plainly acts. Using a simple principle, it achieves all that we observe and more. As I noted before, we are wrongly projecting our own experience on our world. Instead, we should allow our world to project its experience on us. This is the key to the appreciation of the *ground* Gestalt of reality, which is about nature as process. The Swiss biologist, Jean Piaget, may have identified why we have failed to do so thus far. Piaget became known through his theory on the cognitive development of children. [35] He argued that the development of children also resembles that of humankind.[36] Piaget identified four developmental stages. In the third stage, the ability to understand the feelings of another (or empathy) develops. In the fourth stage, when humankind acquires the capacity to relate symbols to abstract concepts, egocentric thinking returns. Needless to say, the development of humanity is beyond the fourth stage of cognitive development. Yet, when Einstein explained the nature of relativity, he still interpreted our world through the eyes of *human* observers. So, what really is nature's view?

Discussion

The examples of Jobs and the Higgs boson may seem frivolous at first sight but they have been carefully selected to illustrate how the extremes of our experience of reality might touch when the latter is viewed as a behavioral phenomenon. Of course, the behavioral nature of fundamental particles is nothing new and has recently been reiterated once again in an article about the wave-like nature of extra large molecules.[37] So, what's new? The crux of the new paradigm of reality is in the explanation of *causality*. What causes the reality that we observe in all its forms and shapes? Effective scientific research may not begin when we have identified the fundamental entities of which the universe (and society!) is composed, as Kuhn had stated, but rather when we have identified the basic phenomenon that "produces" these entities, knowing that no "assembly" takes place in nature. The envisaged paradigm shift moves us away from a common fascination with geometry. The spatial world that seems to surround us is a byproduct of something more basic. Hence, my examination of the Copernican revolution as a reference case seems justified because it involved a paradigm shift that was as basic as the one that we are now facing. What's more, there is a remarkable parallel between the Copernican era and the state of theoretical physics today. Reminiscent of the Ptolemaic model, the complex models used by physicists fairly accurately predict the practical observations. As history has shown, however, that does not mean that the worldview behind these models is true or correct. Then again, what is at stake is not just another technical view of our world. Psychology, even sociology lies at the heart of this, which was a reason to involve the human dimension. We are not just dealing with mere "interpretation" but with "Gestalt": an interpretation of an organized whole that is perceived as more than the sum of its parts.

Notes

---

[1] Walter Isaacson, *Steve Jobs*, Simon & Schuster, New York (2011). In fact, my daughter, Myrthe, and her husband, Bas Kemme, gave this book to me.

[2] Marc van der Erve, *Temporal Leadership*, European Business Review, Vol. 16, No. 6 (2004). My article received a citation of excellence by an independent review board. Many years later, the companies that I analyzed were still on the path of development that I had identified.

[3] Susan Watts, *CERN scientist expects "first glimpse" of Higgs boson*, BBC News (7 December 2011).

[4] Brian Vastag, Joel Achenbach, *Scientists discover new subatomic particle at the center of everything*, The Washington Post (4 July 2012).

[5] *The Higgs Boson: Fantasy turned reality*. The Economist (14 December 2011).

[6] I mean the age of our universe, that is, from the Big Bang until today.

[7] That person was Ad Nollen, a friend and retired orthopedist with an interest in matters that concern the origin of nature.

[8] Albert Einstein, *Relativity: The Special and General Theory* (Revised edition, 1924). Einstein notes: "We entirely shun the vague word "space", of which, we must honestly acknowledge, we cannot form the slightest conception, and we replace it by *motion relative to a practically rigid body of reference*." He then continues to explain that a coordinate system represents such a body of reference. So, physicists essentially adhere to the idea of a world of things too.

[9] Aristotle, *On Physics*: http://classics.mit.edu/Aristotle/physics.mb.txt

[10] I am well aware of Descartes' statement "Cogito ergo sum" (I think, therefore I am), which suggests a spiritual dimension of existence. I will touch upon this further on in this book, particularly when discussing how the behavioral view of our world affects our understanding of the senses.

[11] You'll find a useful summary of the wave-particle duality here: http://en.wikipedia.org/wiki/Wave-particle_duality.

[12] In 1817, Augustin-Jean Fresnel mathematically confirmed the empirical findings of Thomas Young: http://en.wikipedia.org/wiki/Augustin-Jean_Fresnel

[13] Einstein illustrated that energy is equivalent to mass. In certain cases, energy is partially turned into mass. Photons do not have a rest mass. They are all energy.

[14] Arthur Koestler, *The Ghost in the Machine*, Arkana Books, London (1967, 1989), p. 45. You'll find a quick, albeit somewhat biased summary of the holon story here: http://en.wikipedia.org/wiki/Holon_(philosophy)

[15] Thomas Kuhn, *The Structure of Scientific Revolutions*, University of Chicago (Third edition, 1996).

[16] You'll find a neat summary of the model that Ptolemy developed here: http://en.wikipedia.org/wiki/Deferent_and_epicycle

[17] Owen Gingerich, *The Book Nobody Read*, Walker (2004), p. 54.

[18] Kuhn also suggested that unexpected discoveries, such as the discovery of X-rays, might set off a scientific revolution.

[19] Jerome Burner, Leo Postman, *On the Perception of Incongruity: A Paradigm*, Journal of Personality, 18, pp. 206-223 (1949). Here is a link to the article: http://psychclassics.yorku.ca/Bruner/Cards/

[20] Arthur Koestler, *The Act of Creation*, Arkana Penguin Books, London (1964, 1989) p. 234.

[21] I completed the sentence from Aristotle's treatise *On Physics* that I quoted earlier. Here's a link to his essay again: http://classics.mit.edu/Aristotle/physics.mb.txt

[22] Libbrecht used a process chamber in which the saturation of water droplets, electric field, temperature and purity of the water droplets can be controlled. By growing snowflakes at different temperature levels while maintaining the other environmental conditions at a certain level, Libbrecht could accurately evaluate the effect of the temperature on the snowflake structures that emerged.

[23] Elizabeth Weise, *Scientist wants to make snowflake formation crystal clear*, USA Today (18 December 2011). The exact statement of Libbrecht was: "Why snow crystal shapes change so much with temperature remains something of a scientific mystery." Here is the related link to the website of Libbrecht: http://www.its.caltech.edu/~atomic/snowcrystals/primer/primer.htm

[24] Kenneth Libbrecht, *The Physics of Snow Crystals*, Institute of Physics Publishing, Reports on Progress in Physics, 68 (2005) pp. 855-895. Download it here: http://www.its.caltech.edu/~atomic/publist/rpp5_4_R03.pdf

[25] The multiple pictures of snowflakes in the publications of Libbrecht inspired me to refer to his "powerful" focus on the *figure* Gestalt of reality. You'll find the statement quoted on the following page of his website: http://www.its.caltech.edu/~atomic/snowcrystals/faqs/faqs.htm

[26] Kenneth Libbrecht, *Observations of an Edge-enhancing Instability in Snow Crystal Growth near -15 C*, arXiv (11 November 2011): http://arxiv.org/pdf/1111.2786v1.pdf

John Barrett, Harald Garcke, Robert Nürnberg, *Numerical computations of facetted pattern formation in snow crystal growth*, arXiv (6 February 2012): http://arxiv.org/pdf/1202.1272v1.pdf

[27] Ron Cowen, *Snowflakes Re-created Using Physics*, Scientific American (18 March 2012).

[28] Adrian Bejan, *Shape and Structure, from Engineering to Nature*, Cambridge University Press (2000).

[29] Motion is typically referred to as kinetic energy. Kinetic energy comprises translation energy (movement), rotation energy, and vibration energy.

[30] Ilya Prigogine, *Time, Structure, and Fluctuations*, Nobel Lecture (1977). Prigogine's work on minimum entropy production suggests that in many open systems the state of least resistance prevails.

[31] Stephen Wolfram, *A New Kind of Science*, Wolfram Media (2002), pp. 371. Although Wolfram typically focuses on the *figure* Gestalt of reality, he illustrates how a simple add-on process involving a basic shape produces structures that are remarkably similar to real snowflakes.

[32] Further on in this book, I will discuss in detail the crucial idea of "inequality". Inequality (rather than equality) is at the heart of everything that "exists".

[33] This is in accordance with the zeroth law of thermodynamics, which states that nature seeks to establish an equilibrium or equality as soon as an inequality arises. The Second Law suggests that the conversion of energy (into other forms) comes at a cost or fee. This fee involves entropy or disorder, typically the haphazard motion of molecules and particles, a useless form of energy. The American scientist, Rod Swenson, argued that nature produces entropy as fast as the local circumstances allow. Swenson referred to this propensity as the Fourth Law: Rod Swenson, T*he Fourth Law of Thermodynamics or the Law of Maximum Entropy Production*, Chemistry, Vol. 18, Issue 5 (2009). The qualification of the Fourth Law of thermodynamics, however, had earlier been reserved for Lars Onsager's "reciprocal relations", which show the equality of certain ratios between flows and forces in thermodynamic systems out of equilibrium: Richard Wendt, *Simplified Transport Theory for Electrolyte Solutions*, Journal of Chemical Education, Vol. 51, p. 646 (1974). Of course, *Onsager reciprocal relations* may produce the effect that Swenson identified once flows and forces start reinforcing one another.

[34] String theory pictures particles as 1-dimensional things rather than as points in space. It hinges on a law, which says that strings move to minimize their spatial surface not unlike soap bubbles do. This brings us back to the *figure* Gestalt of reality, which revolves around space-time (geometry) and things. Least-energy fitting behavior patterns or inequality border dances do not

necessarily produce the smallest possible surface area. Lee Smolin, *The Trouble with Physics*, Houghton Mifflin, Boston-New York (2006), pp. 103-113.

[35] William Huitt at al., *Piaget's theory of cognitive development*, Educational Psychology Interactive, Valdosta State University (2003). Here is a link to this summary article: http://www.edpsycinteractive.org/topics/cogsys/piaget.html

[36] Jean Piaget, *Genetic Epistemology*, Columbia University Press (1968). This is a speech by Piaget.

[37] Clara Moskowitz, *Quantum mechanics on steroids: Even the largest molecules behave like waves*, MNN (27 March 2012).

Connecting The Dots

In 2005, in a commencement speech at Stanford University, Steve Jobs searched for an explanation of the developments with Apple. "It was impossible to connect the dots looking forward but it was very, very clear looking backward", he said.[1] Then, he added something that summarized in a breath the findings of Thomas Kuhn about the way scientific revolutions come about. It was also the kind of advice you would expect from a true paradigm shifter: "You have to trust that the dots will somehow connect in the future. Believing that the dots will connect down the road will give you the confidence to follow your heart even when it leads you off the well-worn path." Unsurprisingly, Apple revolutionized the personal-computer market again when it introduced the "iPad" 5 years later. To Kuhn, dots simply were anomalies; findings that are either in conflict with accepted ideas or on the edge of what he called "normal science". So here

I summarize the anomalies that I identified. Some appeared to go against the grain of established thinking. Others turned out to have the potential to revolutionize the science that we are familiar with. Together they gradually redirected my view of the world from *figure* to *ground* Gestalt, from *symptom* to *cause*. I was first alerted to these anomalies when I started searching for an explanation of the change in the company that I was working for.[2] At the time, I didn't realize that I was at the beginning of a long and surreal path that led from the usual worries of leaders about the performance of their organization to the odd nature of time and matter. My investigation drifted unavoidably from the societal perspective to questions that kept theoretical physicists up at night. However, I never lost sight of the societal question that I started off with. Every now and then I would return to it to assess the effect of the dots that I had identified. At some point, something odd happened that significantly slowed me down. When trying to explain why organizations change, I discovered that I needed to explain why organizations exist at all. Then, when trying to explain why organizations exist, I discovered that I needed to explain why *anything* exists. Considering such a preposterous challenge, I often wondered what made me say to myself: "Hold on!"[3]

The Great Leadership Deception

When the company that I worked for early in my career (Digital Equipment) started changing from an entrepreneurial stance to one that was more focused on operational control, I searched for similar changes elsewhere in industry. I learned that established corporations, such as General Motors, Kodak, Texas Instruments and Philips, were trying to achieve quite the opposite. By shaking up their bloated organizations and by pushing responsibilities down the management hierarchy, these companies were hoping to revive entrepreneurship, something that we appeared to defocus from after years of rising revenue growth. I made note of these opposite trends in a book with strategies for the development of flexible organizations.[4] I sent a complimentary copy to Wisse Dekker, then CEO of Philips. Some time later, I met a chap who had attended a guest lecture by Dekker at the Delft University of Technology. Apparently, Dekker had advised his audience to read my book. Tongue in cheek, the chap added that Dekker must have done so because he was mentioned in it. Having given the matter some thought, I disagreed. Dekker was mentioned but barely so. Instead, I reasoned, his advice must have been inspired by the belief that leaders can shift the focus of their company at will, a belief that my book seemed to reinforce. In a subsequent book, still in line with that belief, I discussed the approach that I used to help reframe our company both from a business and cultural perspective.[5] I added a most curious connection between the attitudes inside a company and its revenue growth rate. The latter typically traces the shape of a bell as it increases, peaks and, eventually, declines (see Figure 2). The attitudes change in sync. As the growth rate ramps up, a company is more open to influences from outside. The links inside the organization are loosely coupled which allows it to adapt promptly. Once success is firmly repeated, a company becomes more inward focused as it organizes itself to sustain its growth path. Preprogrammed responses to signals from inside and outside the company start populating the collective frame of reference. This causes the links inside the organization to become more tightly coupled and integrated. Whereas the corporate vision inspired the attitudes at first, the corporate culture determines the conduct now. This state continues until the revenue growth rate declines as a result of adverse developments externally and complacency internally. In the end, established procedures are reevaluated and pushed aside to make room for simpler ways. In the process, links are cut and the organization fragments. When, at last, a company seeks guidance from the market in search of profit, it opens up again to influences from outside. This often takes place too late when the company's shareholders have lost patience and are thinking of selling the business. By the time I published my book and in the years that followed, the revenue growth rate of our company started sliding despite the frantic search for strategic answers by the company's leadership and, if I may add, despite my reframing workshops, however well received.[6] Considering the state of our company at the time, its portrayal as "change master" by management guru, Rosabeth Moss Kanter, some seven years before seemed ridiculous, even irritating because of her mechanistic analysis of role models seemingly ignorant of the evolving attitudes in companies that inevitably ride the bell-shaped growth-rate curve.[7] Particularly puzzling was the same denial by hordes of experienced leaders, who bought her book with the belief to find in it the recipe for eternal corporate success. As I'll note later, other management gurus, or should I say "charlatans", continued to reinforce this belief with unbelievable success. In 1992, I left the company to become a consultant and work on a doctorate.[8] On the day I left, I visited the office of the company's European president, Pier-Carlo Falotti, to say goodbye. Under his watch, the European division of the company had grown into a billion-dollar business. He had invited me regularly to evaluate the state of the company, something that I had cherished as a rare privilege. I was sad to hear from him that he too would be leaving the company that day. As I'll explain further on, this event inspired me to reevaluate the role of leaders eventually. In 1998, some six years later, the company, utterly incapable of mastering change, ceased to exist when a competitor absorbed it.[9]

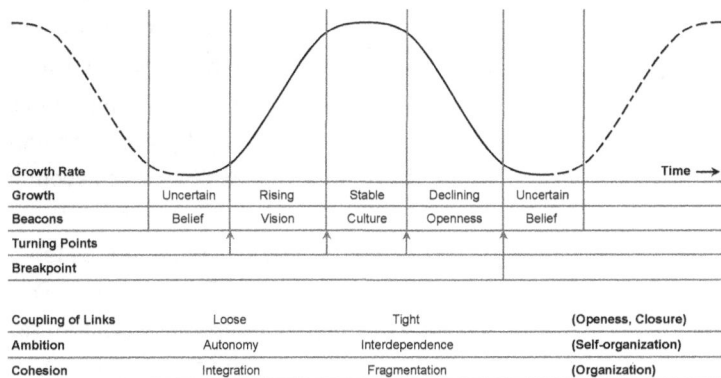

Growth Rate					Time →
Growth	Uncertain	Rising	Stable	Declining	Uncertain
Beacons	Belief	Vision	Culture	Openness	Belief
Turning Points					
Breakpoint					

Coupling of Links	Loose	Tight	(Openess, Closure)
Ambition	Autonomy	Interdependence	(Self-organization)
Cohesion	Integration	Fragmentation	(Organization)

Figure 2 Corporate growth cycle, its turning points and breakpoint.

In the meantime, I continued my investigation into the predictable evolution of corporate attitudes during growth-rate cycles and reported my findings in a doctoral thesis.[10] In it, I evaluated from a sociological viewpoint the idea that companies are finite phenomena. In other words, an organization grows to achieve something after which it may disintegrate and even self-destruct. I assumed that, to a company, achievement means the fulfillment of certain needs of a customer group.[11] I coined the term, "social elasticity", to capture concisely the evolving state of companies. A company resembles an elastic band in that it becomes brittle when it ages. As it traverses successive stages of growth guided by distinctive beacons such as belief, vision, culture, and open-mindedness, it is elastic initially. However, when links inside the organization multiply and become more tightly coupled, it becomes less elastic, even brittle up to a point where it may disintegrate or break when the environmental

39

conditions force it to turn itself from the inside out. The beacons evolve likewise. They are elastic at the outset (from an interpretation point of view) then become inelastic once they have become better defined. Inelastic beacons leave less room for exploration and initiative and, as a result, produce various growth-stage-specific turning points. Only rarely, companies (and their leaders) manage to regain elasticity by remixing and adding corporate components when trying to climb the revenue growth-rate curve again. This may explain why the age of firms in the S&P 500 Index is 15 years on average.[12] Even in the stable post-World War II period "about one-half of the *Fortune Five Hundred* companies turned over within a generation, all either disappearing altogether or dropping out of the front ranks".[13]

In 1994, in the year when my doctoral thesis appeared as a book and despite these lifecycle statistics, management guru, Jim Collins, published *Built to Last*.[14] In his book, Collins reinforced the belief that companies can be built to last forever and, more importantly, that leaders can shift the focus of their company at will if they follow the "right" approach. *Built to Last* remained on the *Business Week* best-seller list for six years and sold in the millions. This meant that Collins's findings, like those of Moss Kanter, were clearly accepted as "established science". My findings represented an anomaly at best.[15] While I continued on what appeared to be a hopeless path of inquiry, increasingly articulating my claim that organizations are not "built to last" but, instead, "grown to achieve", Collins co-authored bestseller after bestseller, banking on what leaders apparently wanted to hear. Relief came 17 years later when *The Economist* wrote, as I had done before, that several of the example companies quoted by Collins "had fallen from grace" (putting it mildly). With the next statement, *The Economist* puts the finger on the pulse.[16]

Mr. Collins might profit from a bit more willingness to admit that,
like all management gurus, he is dealing in clever hunches rather
than built-to-last scientific discoveries.

In 1998, a sequel to my doctoral thesis was published under the title *Resonant Corporations*.[17] The title promised a "how to" book, which it wasn't. Inspired by the book *The Origins of Order* by the American biologist, Stuart Kauffman,[18] I had originally suggested titling it *On the Origin of Corporate Growth*. I now feel that this would have been a better title because I had produced an appraisal of stages of corporate growth, not a silver bullet.[19] In my book, I started off with an analysis of a leadership survey, which I undertook to confirm the consistency of the evolving attitudes in each stage of corporate growth. I also checked how the actual situation related to the one desired by the executives that had contributed to my survey. The results were revealing not just because they confirmed my findings but also because they brought to light a misconception about the way companies evolve. The executives derived the ideal resolve of corporate growth-related issues from the stages of "rising" and "stable" growth no matter which stage their company was traversing. This pointed to a narrow, growth-biased perspective of corporate existence, meaning a denial of the other stages of growth, where different issues and solutions govern. I asked myself why leaders and, for that matter, consultants[20] were persistently ignoring the stages of "declining" and "uncertain" growth, stages that are known to revolve around "creative destruction" and "reinventing the company", as Joseph Schumpeter had said.[21]

The survey participants clearly believed that they could bring back corporate growth (including the upward momentum of their company's share price)[22] by restoring conditions that worked in the previous stage of "stable" growth. In other words, by turning back the clock of corporate evolution. This pointed to fatal flaws in the perception of "organizational reality". First, the reality of organization was equated with and limited to stages of emergence, in which growth-generating behavior patterns of employees emerge, proliferate and endure. Second, obviously dislodged from the realities external to their organization, [23] the survey participants appeared to think that they could restore growth simply by reintroducing (or copying) *internal* conditions that worked in a previous stage. Third, the survey participants appeared to ignore, maybe even deny the cycle of growth stages, the cycle of their company included. Considering these flaws, I wanted another, less-biased source of explanation.

Hoping to shed light on the growth stages that the survey participants had failed to distinguish (probably because they felt that these stages unfolded beyond their control anyway), I embarked on a search for such a source of explanation. This led to an appraisal of "natural" phenomena of organization. I started studying the work of experts who had something to say about the emergence of organization or order. Kauffman, for example, showed that evolution involves fitness landscapes where order emerges when certain conditions come together. But, what kind of order and which conditions? Or, as the world's expert on fossils, Richard Leakey, said: "You can lay out all the fossils that have been collected and establish lineages that even a fool could work out. The question is why, how does it happen?"[24]

The research of the Russian-born Belgian chemist, Ilya Prigogine,[25] focused my attention on an interesting link between chaos and order and how the latter increases the capacity to transmit and convert energy, not unlike what a company does when it turns ideas, materials and labor into products. Apparently, if you heat a thin layer of liquid from below, the orderly flow of molecules in the liquid magically emerges from their chaotic movement. When the inequality between the temperature at the bottom and surface of the liquid reaches a certain level, only an agitation on the surface of the liquid is needed to set off the orderly movement of molecules. The Danish physicist, Per Bak, argued that such an agitation acts like the grain of sand that triggers an avalanche in a pile of sand.[26] It is the straw that breaks the back of chaos. The orderly flow of molecules shows on the surface of the liquid as an organization of hexagonal cells, which resembles a honeycomb. In the middle of each cell, heated molecules rise to the surface where they shed their heat as they move along the surface to the edge of the cell. At the edge, they are ejected to the bottom where they repeat the cycle. But, how does an agitation on the surface of the liquid, produced by some unknown external aberration, bring about orderly flows? And, how does order manage to spread (and win from chaos) in an environment of wildly behaving liquid molecules?

Based on an analysis of continental ice-volume variations over a period of a million years, the Italian scientists, Roberto Benzi, Alfonso Sutura and Angelo Vulpiani discovered that agitations, produced by the Earth's rotational irregularities, had caused its climate to swing in an orderly way between long cold and warm eras. Short-term climate changes as a result of annual variations in solar radiation had amplified these agitations.[27] In fact, Benzi, Sutura and Vulpiani demonstrated that a weak signal (rotational irregularities) had somehow resonated with chaos (short-term climate changes), which in turn had pushed it over a threshold so that it spread and triggered orderly climate behavior around the globe (long cold and warm eras). This meant that chaos was not just the opposite of order but also a vital ingredient in the recipe that produced it. To put it differently, chaos was the soil that nurtured order. Some years later, I learned about the research of the Japanese scientist, Ichiro Tsuda, who had arrived at the same conclusion when it comes to the functioning of the brain. [28] Our brain is in a state of constant chaos as a result of electrical currents that are randomly leaking from trillions of neuron connections. Then again, our thinking hinges on orderly behavior involving clusters of widely spread neurons that produce distinct sequences of electrical pulses all at the same time. Apparently, when a weak signal from the senses (or inside the brain) resonates with the noise of trillions of leaking neuron connections, it is pushed over a threshold so that it can spread and trigger orderly neuron-cluster behavior across the brain. Tsuda noted that the same principle explains the functioning of other biological networks, such as immune networks, metabolic networks, and the network of capillary blood vessels, and even the functioning of societal networks.

In *Resonant Corporations*, I tentatively referred to these naturally emerging phenomena of order to make sense of the two largely neglected stages of growth. In a stage of declining growth, an organization fragments until it meets a breakpoint at which it might dissolve back into the sea of societal noise from which it once emerged. In a stage of uncertain growth, that same sea may nurture new forms of organization at places where a niche of customer needs deepens to a certain depth. At those places, an agitation, such as the call of an entrepreneur, acts as the grain that triggers an avalanche of orderly human behavior patterns, patterns that try to fulfill these needs.

Some were quick to dismiss this parallel, saying that people weren't molecules.[29] Of course, not many knew that the Austrian physicist and inventor of quantum mechanics, Erwin Schrödinger, had already suggested a parallel between a flame and the human body. Different in complexity, both essentially are fleeting streams of order that involve oxidizing molecules.[30] Yet, at that point, I realized that I couldn't push the envelope by metaphors alone. The answer to Leakey's question "Why, how does it happen?" was still missing. Moreover, the interpretation of the reality of our world at the time left too big a gap between the different "corpuscles" that it supposedly contained. Tired of my contrarian role, I left these two questions unanswered for nearly a decade.

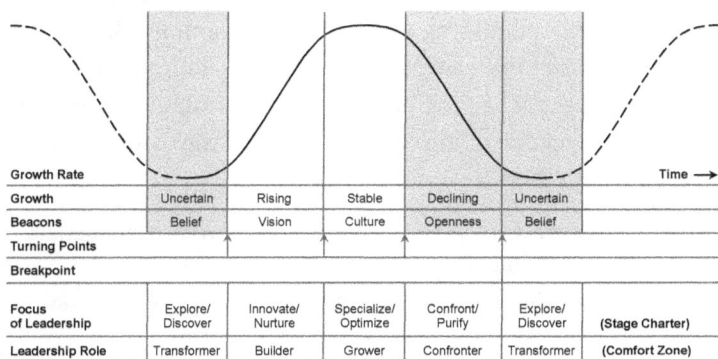

Growth Rate						Time →
Growth	Uncertain	Rising	Stable	Declining	Uncertain	
Beacons	Belief	Vision	Culture	Openness	Belief	
Turning Points						
Breakpoint						
Focus of Leadership	Explore/ Discover	Innovate/ Nurture	Specialize/ Optimize	Confront/ Purify	Explore/ Discover	(Stage Charter)
Leadership Role	Transformer	Builder	Grower	Confronter	Transformer	(Comfort Zone)

Figure 3 The largely neglected stages of growth and leadership roles.

In the mean time, I focused on organizational challenges and leadership needs as advisor. This gave me the unexpected opportunity to interview many top-level executives. It did not take me long before I started relating their careers and comfort zones to the stages of corporate growth that I had identified. I came to appreciate better the role of Pier-Carlo Falotti at Digital Equipment during its stage of stable growth. I also learned that an evolving organization often forces leaders to function in a role that may not be their natural one. This left a largely unused opportunity to identify leaders whose comfort zone would better fit the needs of an evolving organization (or nation).

Executives generally perceived a comfort-zone appraisal as more imaginative than a traditional psychological assessment because it helped them identify in which stage of corporate growth they might function most effectively. [31] On the other hand, some corporate leaders resisted the idea that their reign might ideally have to become limited to one stage of corporate growth only, understandably, particularly by those who were in power. At some point, a befriended executive [32] suggested writing up my findings. That set me off on another lengthy path of inquiry. After publishing a paper with the unmistakable title *Temporal Leadership*,[33] I was determined to find an answer to the questions that I had left unanswered for so long and to leave no stone unturned in doing so. I started from the usual premises by Aristotle and Descartes.

> *Our world consists of substances, which are divisible into other substances, if not substances then attributes. (Aristotle)*
> *Our world consists of corpuscles of matter that can be divided and assembled into other corpuscles. (René Descartes)*

However, in my mind, I was ready to distance myself from these if needed. Following Leakey, one question guided me:

> *What really shapes our world (why, how does it happen)?*

Having veered off into the domain of "natural" phenomena of organization [34] (in search of an explanation of the two neglected stages of growth), my inquiry was no longer limited to human forms of organization. In what follows, I'll discuss the alleys that I explored to find dots that might spontaneously connect and provide me with the answers that I was looking for. In these alleys, I found clues, anomalies and even generally accepted misconceptions that, in the end, led to a different perspective of what truly shapes reality, even in business.

The Alley Of Natural Growth And Selection

In 1789, the British economist and demographer, Thomas Malthus, published anonymously the first version of his *Essay on the Principle of Population* stating that the latter *"affects the future improvement of society"*.[35] Alarmed by the dramatically increasing life expectancy of children that followed the industrial revolution, Malthus described the dismal prospects for Great Britain if it failed to curb population growth. "The increase of population is necessarily limited by the means of subsistence" such as the food supply, he wrote. If population growth is left unchecked then "the population is kept equal to the means of subsistence, by misery and vice". Malthus also opposed the prevalent belief in Europe at the time that society could be improved almost limitlessly. He particularly criticized the views of his countryman, the novelist, William Godwin, and those of the French mathematician, Nicolas de Condorcet. To Malthus, the solution was restraint such as birth control (a strategy which China adopted later). Influenced by the French revolution, however, Godwin believed that the state should not be trusted to do what is right for the community.[36] De Condorcet instead argued that societal improvement would follow "the progress of the human mind" eventually.[37] The views of Malthus produced a breakthrough in the minds of two scientists.

In 1838, the Belgian mathematician, Pierre-Francois Verhulst, introduced an equation that depicted an S-shaped population-growth curve (Figure 4), which allowed for a limited means of subsistence.[38] Following the shape of this "S-curve", growth increases slowly at first then rises steeply before tapering off and flattening when the limit of the means of subsistence is reached. Verhulst's S-curve illustrated both the development and frontier of population growth and confirmed

Malthus's gloomy forecast. Interestingly, at the time of Malthus, the population of Great Britain was about 7 million people. Sixty years later, it had doubled and would double again in the fifty years that followed.[39] What had happened?

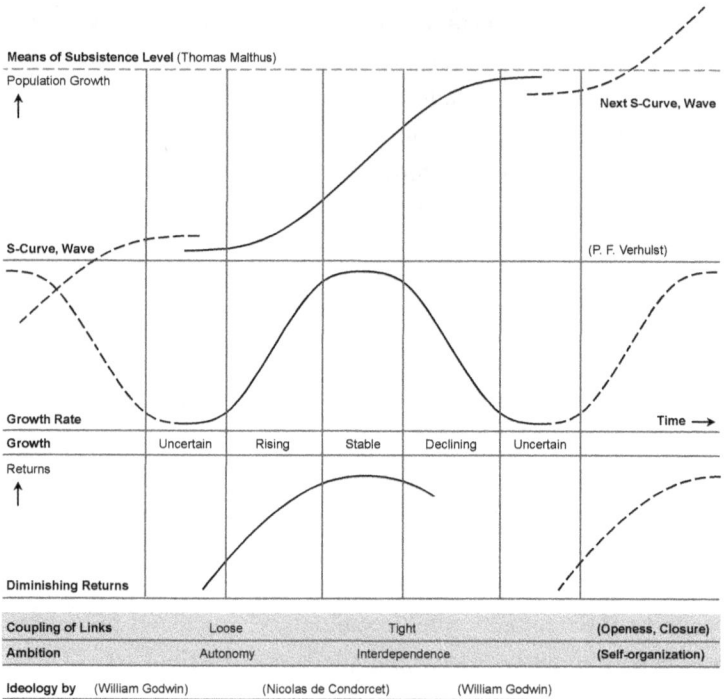

Means of Subsistence Level (Thomas Malthus)

Population Growth ↑

Next S-Curve, Wave

S-Curve, Wave

(P. F. Verhulst)

Growth Rate   Time →

Growth	Uncertain	Rising	Stable	Declining	Uncertain

Returns ↑

Diminishing Returns

Coupling of Links	Loose	Tight		(Openess, Closure)
Ambition	Autonomy	Interdependence		(Self-organization)
Ideology by	(William Godwin)	(Nicolas de Condorcet)	(William Godwin)	

Figure 4 In a stable environment, society may ride successive S-curves.

As a result of technology- and administration-driven (and other) productivity increases, the means of subsistence had no doubt increased through successive waves of improvement, each wave resolving a temporary limit and producing the next population-growth curve or S-curve. With reference to my own findings, the attitudes in society must have changed in sync each time such a wave unfolded. Thus, at the end of a wave, in line with Godwin's argument, citizens had to shake off an inhibiting legacy of established societal links for new ideas to develop. Then, as Malthus had suggested, a period of restraint created time and space for such ideas to materialize. Eventually, along the lines of Condorcet's views, human thinking advanced once these ideas had become rooted in society and helped increase its means of subsistence. Therefore, despite Malthus's critique, Godwin, Condorcet and even Malthus himself had essentially sketched the situation at different stages of societal emergence (including those that my survey participants appear to have ignored).[40]

By now, researchers have refined and expanded the work of Malthus. First, they highlighted some obvious flaws, such as Malthus's assumption of a stable external environment and the missing effects of competition.[41] Of course, natural disaster or the arrival of a party that competes for the same resources may severely disrupt and even permanently destroy the means of subsistence. Second, they identified similar phenomena of growth in practically every field, from epidemics and product revenues to the use of energy sources.[42] In sum, no matter whether it concerns corporations, neural networks, or any other form of organization, the S-curve represents a universal image of reality emerging, a true fractal of "natural growth".

The S-curve also prompted an interest in a related curve that shows how the evolving complexity of a growing concern affects its capacity to produce additional returns. From very early on in human history, farmers were aware that the increase in the net number of seeds harvested gradually diminished as new seeds were planted. This is why they shied away from fully utilizing cropland.[43] Each new seed not only faces competition for moisture, minerals and sunlight from earlier planted seeds but it also demands a proportionally bigger share of the labor needed for harvest due to the absolute quantity involved.[44] In business, the division of labor is known to increase productivity because it fosters specialization. However, when you divide a manufacturing task into subtasks for this reason, you increase the need for overhead, for example, to balance the workflow, to control product quality, and to manage suppliers. As a company grows, the continued division of labor and the resulting need for overhead gradually reduces its capacity to generate additional returns. As a result, the net increase of returns diminishes as a company progresses along its growth curve. So, developing in sync with the S-curve, the curve of (diminishing) returns (Figure 4) represents another powerful image of reality emerging.

I noticed that scientists and consultants alike often ignore the link between the curve of diminishing returns and the S-curve.[45] Evaluating the former in isolation, they assume that the clock of corporate and societal emergence can be turned back once the increase in returns turns negative. By the introduction of some form of strategy, they hope to create the conditions needed to climb up the curve all the way *back* to its maximum.[46] Considering the path of societal and corporate emergence that runs *forward* in time along the S-curve, this is like wanting to bring back a lump of sugar after it has dissolved in a cup of tea.[47] Higher returns may only be brought back in a subsequent cycle of growth, which involves the earlier-discussed transition of attitudes along the thinking of Godwin, Malthus, and Condorcet.

In conclusion, the S-shaped curve of natural growth and the curve of diminishing returns, that unfolds in sync, are robust images of reality emerging. However, they did not provide an answer to my guiding question "What really shapes reality?" In other words, they failed to tell me "why or how it happens". The

S-curve and the curve of diminishing returns essentially are *attributes* of the process of emergence only.

The work of Malthus also produced a breakthrough in the mind of the British naturalist, Alfred Russel Wallace. In 1844, at the age of 21, Wallace read the sixth edition of Malthus's essay *On the Principle of Population*. In the same period, he also read Charles Darwin's popular *Journal of a Naturalist's Voyage on the Beagle*, which inspired him most. Four years later, after a short-lived career in architecture and civil engineering, he followed in the footsteps of Darwin joining a naturalist's expedition to the Amazon region and Rio Negro area. Eventually, in 1855, one year into a 7-year expedition to the Malay Archipelago, Wallace wrote a paper, in which he shared the observation that "every species has come into existence coincident both in space and time with a pre-existing, closely allied species. All species together thus formed a branching tree."[48] Yet, at that stage, neither Wallace nor Darwin knew what made this tree grow.[49] They didn't know "why, how it happened". One day, on an island not far from New Guinea, Wallace was forced to rest "muffled in blankets in the cold fit of an attack of intermittent fever". In this dismal state, he suddenly remembered reading the gloomy essay of Malthus and a more constructive reading of Malthus's dark analysis dawned on him. "The struggle for existence," of which Malthus spoke, could also be seen as a process that "would necessarily improve the race". With reference to the shortage of food and other means of subsistence, which Malthus had identified as causes that would curb population growth, Wallace explained the line of discovery in his autobiography.[50]

It occurred to me that these causes are continually acting in the case of animals also; and as animals usually breed much more quickly than does mankind, the destruction every year from these causes must be enormous in order to keep down the numbers of each species, since evidently they do not increase regularly from year to year, as otherwise the world would long ago have been crowded with those that breed most quickly. Vaguely thinking over the enormous and constant destruction, which this implied, it occurred to me to ask the question, why do some die and some live? And the answer was clearly, on the whole, the best fitted live.

So, in an era, the population-growth curve of a variety of a species that best fitted the environmental conditions would rise while the curves of varieties that fitted less would taper off and flatten. Wallace noted that "even the peculiar colors of insects, so closely resembling the soil, leaves, or trunks on which they reside" determined the fit. "Varieties of many tints may have occurred. Yet, those races having colors best adapted to concealment from their enemies would survive the longest." Wallace realized that he had found "the long-sought-for law of nature that solved the problem of the origin of the species" and explained how the branching tree of species had grown. Within days, he had sent a letter to Darwin who would refer to this law as the law of "natural selection" in his epic book, *On the Origin of Species*.[51] As early as 1855,[52] Wallace also hinted that not just the physical and behavioral traits of individuals but also the behavioral traits of groups contributed as variables to the equation that determined the best-fitted variety. [53]

The comparative abundance or scarcity of the individuals of species is entirely due to their organization and resulting habits.

In 2012, the American evolutionary biologist, Edward Wilson, took this idea to its extreme. In *The Social Conquest of Earth*, [54] he argued that the survival of the fittest does not just depend on "individual selection" where an individual is willing to risk his survival to save members of his gene pool or kin but also on "group selection". In other words, no matter whether it concerns ants, bees or humans, "each member of a society possesses genes whose products are targeted by individual selection and genes targeted by group selection," and, as Wilson emphasized, the latter are vital. As one reviewer put it, "selfish individuals might beat altruistic ones but groups of altruists beat groups of selfish people".[55] Wilson's views triggered a hot debate. According to the British evolutionary biologist, Richard Dawkins, Wilson's book contained "many pages of erroneous and downright perverse misunderstandings of evolutionary theory".[56] In line with the views by Dawkins, another reviewer wrote: "it's long been known that unrelated individuals can benefit from repeated cooperation with one another, so long as there are mechanisms in place to encourage reciprocity and punish betrayal," (probably referring to the book, *The Origin of Virtue*, by Matt Ridley).[57]

In one of the above reviews, the German evolutionary scientist, Georgy Koentges, summed up the issue with natural selection: "the central problem is the impossibility of defining fitness, whether in organisms, organs, cells, genes or even gene regulatory DNA regions." As I wrote elsewhere, long before the debate that Wilson's ideas set off, the analyses by Dawkins (and now those by Wilson) were well researched and exceptionally detailed. No doubt, the intention must have been to illustrate "the enduring process of tradeoffs" that nature makes. However, "apart from chance [as determining factor] and the overarching principle of the survival of the fittest, each case of species development stands on its own and requires extensive analysis, if not speculation as to why nature favored certain traits over others." So far, "no central idea or process has surfaced that [unequivocally] explains the dynamics of the emerging realities in biological, cultural, and physical fields." [58]

In short, I came to realize that the alley of natural growth and selection, however promising at first sight, had a dead end too. The idea of natural selection did not evoke a precise-enough answer to the question, "How does it happen?" When it comes to the explanation of what shapes our world, natural selection or, for that matter, the survival of the fittest is an attribute of the process of emergence only.

The Alley Of Self-Organization

On my path of inquiry, the ideas of the British-born American anthropologist, Gregory Bateson, probably influenced me most. In the late eighties, I learned that he had used a double feedback loop to explain certain human interactions. As a result, he was recognized for helping to extend cybernetics or systems theory to the social sciences. This reminded me of my engineering thesis, which involved cybernetics too.[59] So, I bought his book, *Mind and Nature*, and (later) its predecessor, *Steps to An Ecology of Mind*, the former, as Bateson noted, "including all that were acknowledged in the preface of the latter".[60] Frankly, I did not find in these books what I expected. I found something much more valuable.[61] In an entertaining account of two lectures (one to art students and one to young psychiatrists, both groups of whom he found to be rather uninformed and even skeptical), he explained how he had brought two brown paper bags, one with a freshly cooked crab and the other with a beautiful large spiral shell. He first showed the crab and asked his audience "to produce arguments that would convince [him] that this object was a remains of a living thing". He added that his audience had to imagine that they were Martians. As Martians they would be familiar with living things (after all, they were living creatures themselves) but they would never have seen crabs. The idea was to distill what distinguishes living things. First of all, his students noticed that the object was "symmetrical", the right side resembling the left. Next, although one claw was slightly bigger than the other, someone observed that both claws were made of the same parts. This was the pattern that connected both sides of the crab. Upon further scrutiny, they "recognized in every leg pieces that corresponded to the pieces in the claw". This was evidence of theme variation or "modulated repetition". Claws and legs were related to one another as much as fingers were to toes, and arms were to legs. Bateson summarized the intermediate finding of his students as follows.

The anatomy of the crab is repetitive and rhythmical. It is, like music, repetitive with modulation.

Next, he showed them the spiral shell and repeated the question: What would convince them that this object had been a living thing? With the ideas of symmetry, repetition of parts and modulated repetition still fresh in mind, the students felt disoriented. A spiral shell is neither symmetrical nor segmented. "It's a figure that retains its shape as it grows by addition at the open end." As Bateson insisted, "his students had to discover [as I had to] that all symmetry and segmentation are a payoff from growth and that growth makes its formal demands and that one of these demands is satisfied by spiral form." In all, Bateson inspired me to search for a *pattern that connects*. The ideas of *symmetry, repetition, modulated repetition, and growth* would play a role in it. Bateson added one more clue:

The right way to begin to think about the pattern that connects is to think of it as a dance of interacting parts.

Even before my doctoral research, I became fascinated with the idea of "self-organization" through Gilbert Probst, a young professor at the University of Geneva, whom I had the pleasure to meet eventually.[62] Probst and his co-editor, Hans

Ulrich, had bundled a collection of articles (including their own) presented at a colloquium on the topic by renowned experts in this field.[63] To paraphrase some recent researchers,[64]

Self-organization is a spontaneous process in which a pattern or shape emerges exclusively from numerous interactions between components at lower-levels, interactions that have been triggered by conditions at a local level and without regard for the emerging shape of the whole.

So, the shape of a spiral shell spontaneously emerges from repeated interactions modulated by local conditions. The shape of an organization spontaneously emerges from repeated human interactions modulated by local (market, resource, and strategic) conditions. The shape of our bodily organs and limbs spontaneously emerges from repeated interactions involving a genetic code that is modulated by local conditions.[65] The shape of our thoughts spontaneously emerges from repeated neuron-cluster interactions that are modulated by local conditions.[66] Ultimately, without the need for a grand design, creation arises spontaneously from repeated interactions that are modulated at a local level. Of course, the question is "how does it happen?"

Through their book, Probst and Ulrich parachuted me right into what appeared to be a most promising field of explanation. One of the contributors to the book was the German physicist, Hermann Haken. Haken had established a new interdisciplinary field of research focused on the study of self-organizing phenomena, calling it "synergetics". Synergetics (which means "working together") explores material and non-material cases of self-organization that arise from the repeated interactions between multiple parts or things.[67] By now, some eighty volumes with experimental and theoretical findings have been published with numerous applications in both the natural and social sciences.[68] The evidence is overwhelming. No matter what things are involved (anything from particles and molecules to animals and people), orderly patterns emerge from chaotic behavior under certain conditions. These conditions typically involve an energy inequality of some kind (thus, a state far from equilibrium). So, in a thin layer of liquid, orderly patterns of molecule behavior emerge only when an inequality between the temperature at the bottom and surface reaches a certain level. In a company, new orderly patterns of human behavior emerge only when an inequality between the actual and envisioned state reaches a certain level.[69]

In 1977, the Russian-born Belgian chemist, Ilya Prigogine, received the Nobel Prize in Chemistry not only for confirming the inequality-inspired emergence of orderly behavior patterns but also for taking this understanding further. First, Prigogine showed that the emergence of orderly patterns involved the conversion of one form of energy into another. For example, in a thin layer of liquid that shows orderly patterns of molecule behavior, heat is converted into motion energy (of the moving molecules) at the bottom. At the surface, molecules shed or dissipate this energy to the atmosphere, at which time energy is converted again. One might argue that a similar conversion of energy takes place in human organizations where ideas, resources and labor are converted into products. Part of the energy used in the process is not recoverable and dissipated or wasted to the environment. Of course, phenomena, that absorb energy at one end to waste it at another, are "open systems". Second, Prigogine observed that the transfer of energy becomes more efficient once orderly behavior patterns have developed (because the latter follow paths of least resistance). Predictably, companies also justify their existence by bridging supply and demand more efficiently. Third, at the time, many scientists believed that any process in nature could simply be reversed.

Prigogine proved this to be too mechanistic a view of nature. The orderly behavior of molecules in a thin layer of liquid can be reestablished by reheating the liquid from underneath but the resulting patterns of orderly behavior visible on the surface as hexagonal (convection) cells would be distributed differently (in part due to the differences in the chance-driven agitations on the surface of the liquid that trigger the emergence of orderly behavior in the first place). Also, the energy that had been wasted before cannot be used this time around. In other words, nature and creation only move forward in time through successive cycles, in which chaos and order alternate, cycles that typically display an S-shaped history of development. Prigogine put the consequences as follows: [70]

> *Irreversibility changes our view of nature. The future is no longer given. Our world is a world of continuous construction ruled by probabilistic laws and no longer a kind of automation. Our world is not a world of being but a world of becoming.*

One of the contributors to the book edited by Probst and Ulrich was the Chilean biologist, Francisco Varela. In his essay, Varela discussed the peculiar behavior of the brain.[71] As a self-organizing phenomenon, the brain does not straightforwardly react to signals received from the senses or elsewhere. Rather, it allows these signals to function as agitations that modulate the thought processes it maintains. So, as another researcher had identified, the brain, while "open" energy-conversion wise, was "closed" from an operational and informational point of view.[72] Yet, "closure" also made the brain more coherent, Varela added, behaviorally, that is. The coherence of its internal behaviors determined its identity. This reminded me of an organization again. At some stage during a revenue growth cycle, a company inevitably becomes more "closed" operationally when it spins a network of preprogrammed (and, thus, coherent) responses to signals from inside and outside the company (hoping to repeat early success more efficiently). During such times, its culture and identity become indeed more distinctive. Varela suggested capturing the brain's ability to "self-produce" its own identity by the term "autopoiesis", which means "self-production". In 1972, Varela and his teacher, Humberto Maturana, had invented the term when debating "what might characterize living systems as unities".[73] Varela's essay was intriguing. The term "autopoiesis" and the ideas of "closure" and "coherence" radiated the promise of a breakthrough. Hence, I read *The Tree of Knowledge* by Varela and Maturana, hoping to find what I was looking for.[74] Instead, I found a rather poetic and, at times, even confusing discussion of cases and conditions that hinted at autopoiesis, a discussion that, in the end, failed to answer the question "Why, how does it happen?" Then again, in his essay, Varela offered an idea that further shaped my path of inquiry: "In principle, there are no self-organizing systems, only self-organizing behaviors." So, might "behaviors" shape our world?

The American evolutionary and systems theorist, Rod Swenson, took public issue with some of the imaginative statements employed by Varela and Maturana in their book.[75]

Not only does [autopoiesis] add nothing to the explication or understanding of spontaneous ordering but it obfuscates such an effort with obscurantist metaphysical baggage which, when unpacked, reveals a set of ontological claims and assumptions not merely so unfounded in fact as to be absurd.

His explicit comments were rooted in the fact that he had found a much less ambiguous explanation. While investigating the work of Prigogine and other researchers, Swenson had come to an exceptional realization that hinted how the spontaneous ordering of parts might come about, a realization that involved the second law of thermodynamics. This law predicts that part of the energy, that is converted or transformed from one form into another through orderly behavior, is wasted. This wasted energy (the cost of transformation, as it were) is so scattered that it cannot be recovered.[76] Scattered as it is, it is in a state of equilibrium and can therefore no longer be used to establish some other inequality[77] (a state far from equilibrium) to induce orderly behavior elsewhere. In 1865, the German physicist, Rudolf Clausius referred to this wasted quantity of energy as "entropy", which, derived from Greek, means "transformation". As Swenson emphasized, the second law of thermodynamics goes further: Nature seeks to maximize the creation of entropy by minimizing an energy inequality *as fast as the local conditions allow*.[78] In other words, nature nurtures the orderly behavior of "components", such as molecules and people, in order to transfer energy more efficiently from places with a relative surplus to places with a relative shortage until equilibrium has been achieved. Basically, this means that orderly behavior is evidence of nature trying to minimize some inequality. Thus, inequality is at the heart of phenomena that display orderly behavior; no orderly behavior can exist without it. Swenson concluded that evidence of orderly behavior at macro levels (such as the distribution of hexagonal convection cells visible on the surface of a thin layer of liquid that is heated from below) emerges from "circular relations" at micro levels where inequality is minimized in unison. The minimization or "breakdown" of inequalities at macro and micro levels inspired Swenson to call these phenomena: "autocatakinetic systems". The term "autocatakinetic" combines the words: "self" (auto), "down" (cata) and "kinetic", the latter originating from "kinein", which denotes "to cause to move".

> *"No claim is made concerning components; in fact it is purposely avoided. Thus dust devils, tornados, convection cells, bacteria,*

ecosystems, civilizations, and the global Earth system as a whole
are all examples of autocatakinetic systems."

So, the non-living and living are autocatakinetic systems. As Swenson noted, "Non-living systems are slaves to their local [inequalities] while the living are not." The display of orderly behavior in non-living systems, such as the convection cells on the surface of a thin layer of liquid (that is heated from below), fades as soon as the heat is turned off. Thus, the fate of non-living phenomena of orderly behavior is directly linked to the fate of the inequalities that sustain it. Living systems distinguish themselves by their ability to search for new sources of (energy) inequality. So, bacteria change direction in search of nurturing streams and human organizations change direction in search of nurturing markets. In sum, by reinterpreting the second law of thermodynamics, Swenson formulated a compelling macro-level response to the question: "What shapes our world?" Our world involves patterns of orderly component behavior that emerge to minimize inequalities (or maximize entropy) as fast as the local conditions allow. When Swenson recalled that nature favors patterns with the highest rate of inequality minimization (or entropy production), he effectively redefined the idea of natural selection and extended it beyond the evolution of living species. Two issues remain. First, Swenson's idea still hinges on "components". But, what are the components at the lowest level? Second, when orderly behavior comes about, how exactly does nature make its trade-offs?

The American biologist, Lynn Margulis, referred to Swenson's work in the introduction of her book *"What is Life?"* as follows. [79]

[Swenson's] universe is pocked by local regions of intense
ordering, including life, because it is through ordered, dissipative
systems that the rate of entropy production in the universe is

maximized. The more life in the universe, the faster that various forms of energy are degraded into heat. Swenson's view shows how life's seeming purpose (its seeking behavior, its directedness) is related to the behavior of heat.

In 1967, Margulis provided microbiological evidence of how bacteria grew more sophisticated capabilities to minimize a broader range of inequalities.[80] In short, some bacteria engulfed other bacteria as predators and became dependent on the capabilities of the engulfed when the latter survived the process. One such engulfing bacterium is the Hatena, which has a whip-like appendage that enables it to swim (flagellum). It typically acts like a predator. But, when it ingests a green alga, it changes from predator to host and switches to photosynthetic nutrition. In the process, the Hatena loses its original feeding device yet acquires the ability to move towards light.[81] Margulis's findings confirmed an evolutionary process called "endosymbiosis"; the term combines the words "within" (endon) and "living" (biosis). Endosymbiosis is more common than we realize. Companies regularly engulf other companies to thrive on the capability of the engulfed. In 2010, Apple "engulfed" the company, Siri, which had developed an App under the name Siri. The App acted as an intelligent personal assistant and was capable of answering spoken questions. First included in the iPhone 4S in 2011, Siri helped Apple move to a new nurturing stream in the market. Justifiably, Margulis and her daughter, Dorion Sagan, argued that evolution is less a matter of competition, as Darwin had suggested, and more a matter of working together.[82] Nearing the end of *"What is Life?"* Margulis defined "life" quite in line with the views of Swenson: "Life is: *moving, transmigrator of matter, transmutation of energy, incessant heat-dissipating chemistry.*" She added something that, as I'll show later, would further inspire me: "Life is: *memory in action.*"

62

In 1996, the Romanian-born American engineer, Adrian Bejan, reworded the findings of Prigogine (*our world is a world of continuous construction*) and Swenson (*nature favors patterns with the highest rate of inequality minimization and, thus, a state of least resistance*) to fabricate another flavor of the second law of thermodynamics. His so-called "constructal law" dictates: "For a finite-size system to persist in time, it must evolve in such a way that it provides easier access to the imposed currents that flow through it."[83] Bejan's version, although puzzling (due to the link between "flows" and "construction"),[84] confirmed that the remarkable findings of Prigogine and Swenson had also reached the arena of engineers. Yet, macro-level as it is, it failed to answer my guiding question, "Why, how does it happen?"

In sum, the idea of self-organization produced powerful clues about what might shape our world. Bateson wrote that it might be "a dance of interacting parts". Haken and Margulis stressed that our world is shaped by components "working together". Prigogine concluded that the clock of nature cannot be turned back and that we live "not in a world of being but in a world of becoming". Varela hinted that behaviors rather than components shape our world. According to Swenson, our world involves patterns of orderly component behavior that emerge to minimize inequalities as fast as the local conditions allow. All the researchers that I evaluated here coined a catching term to sum up their findings:

- "Patterns that connect" (Bateson),
- "Synergetics" (Haken),
- "Dissipative systems" (Prigogine),
- "Autopoiesis" (Varela),
- "Autocatakinetics" (Swenson),
- "Endosymbiosis" (Margulis),
- "Constructal law" (Bejan).

Overlapping as they are, these terms failed to answer my guiding question "How (exactly) does it happen?" These imaginative terms truly deepened my grasp of reality but they are only attributes of the process that shapes our world.

The Alley Of Complex Systems

After venturing through the domains of management theory, natural growth and self-organization, I searched for another perspective of the four stages of organizational growth, hoping to find scientific support for the stage characteristics that I had identified. I wanted to take a more distant view of the matter and had already started referring to natural phenomena as stage metaphors for the two neglected stages, albeit superficially.

In a stage of declining growth, an organization fragments until it meets a breakpoint at which it might dissolve back into the sea of societal noise from which it once emerged.

With reference to Prigogine's work, a stage of declining growth reminded me of the stage of "chaotic" molecule movement, from which orderly behavior might arise eventually. Indeed, if the conditions were right, the next stage would follow.

In a stage of uncertain growth, that same sea may nurture new forms of organization at places where a niche of customer needs deepens to a certain depth.

At that point, consistent with the findings of the Italian scientists, Benzi, Sutura and Vulpiani, and the Danish physicist, Bak, "an agitation, such as the call of an entrepreneur, may act as the grain that triggers an avalanche of orderly human behavior patterns, patterns that try fulfill these needs".

The idea of chaos as precursor to a new cycle of corporate growth was not new to the leadership community. In *Thriving on Chaos*, for example, the American management guru, Tom

Peters,[85] had already spelled out what it takes to identify new market niches. In short, executives should anticipate as well as respond to accelerating change in the market place by focusing on matters, such as innovation. His "keep-it-simple, unduly regimented and even impracticable advisories" (as one reviewer put it)[86] may have lacked finesse but, as I shall explain, the title *Thriving on Chaos* was spot on.

As the investigations of Prigogine, Benzi, Sutura, Vulpiani, and others had illustrated, forms of organization (or orderly behavior) emerge from "chaos". So, when searching the Internet for relevant publications, I kept my eyes trained on this. One day, while jumping from reference to reference, I came across an article by the Japanese mathematician, Ichiro Tsuda.[87] Tsuda, a prominent scientist,[88] whom I had the privilege of consulting a few times later, is a member of the self-proclaimed "Gang of Five", a group of Japanese scientists that has advanced the study of "complex systems" comprehensively.

In complex systems, the overall behavior emerges from a many-to-many relationship involving both the intentional and unintentional interactions of components or parts. A typical example of a complex system is the human brain with a hundred billion neurons that are connected in trillions of ways. Other examples are organizations (people), ecosystems (species) and immune networks (antibodies). Yet, as Tsuda and his colleagues demonstrated, even a growing ensemble of just a few players (neurons, dividing cells, people) soon fits the description.[89] The research by the Gang of Five expanded my understanding of phenomena of orderly behavior (and, thus, organization) in five important ways.

First, it emphasized that the coalescence of component behavior rather than the components themselves determine the identity of a complex system. This represented a shift from a material to a behavioral perspective, a realization that began to dominate my thinking.

Second, it showed that chaos (the ocean of intentional and unintentional component interactions) is not a condition that influences one stage only. Rather, chaos plays a crucial role throughout all stages of orderly behavior development. As Benzi, Sutura and Vulpiani had unveiled, a faint signal, whether intentional or unintentional, may resonate with chaos to such an extent that it is pushed over successive energy thresholds and transmitted across distances.[90] While skating on the surface of chaos, a signal induces orderly behavior along its path. In sum, as "information channel", chaos facilitates the emergence of widely spread orderly behavior as soon as new signals arrive. As such, it helps fireflies sync their flashing, crickets sync their chirping and frogs sync their croaking.[91]

Third, orderly behavior involves synchronization, that is, the simultaneous unfolding of behavior patterns. So, the four stages of organizational emergence that I had identified were also stages of "simultaneity development". I'll discuss later why this realization led me to explore the idea of time.

Fourth, the research by the Gang of Five confirmed that behavior patterns emerge to minimize local inequalities.[92] In the process, behavior patterns mutate until the best performing patterns out-reproduce the less performing ones. To observers, it seems as if successive behavior patterns are "attracted" to a best performing pattern. This explains its somewhat misleading label of "behavioral attractor".[93]

Fifth, Tsuda and his colleagues distinguished themselves by identifying memory as a dynamic and natural phenomenon. They discovered that a faint signal typically first evokes a series of old, unstable behavioral attractors, called "attractor ruins", each such attractor representing a momentary recall.[94] Because the journey along old and unstable behavioral attractors is steadfastly repeated each time when a new signal arrives, the journey itself appeared to behave as an "attractor". The Gang of Five coined the term "itinerant attractor" (travelling attractor) to describe this extraordinary phenomenon. Because itinerant attractors can be observed across all complex systems, the emergence of such a dynamic form of memory is wholly natural. What's more, after evoking a series of temporary old behavioral attractors (momentary recalls), travelling signals usually induce a new behavioral attractor (the current state). Then, by some cascading process, a sense of distinction, if not past and present emerges. Both are at the root of recognition. As the British philosopher, Alfred North Whitehead, said, "Recognition is an awareness of sameness" (and, thus, repetition).[95]

With reference to the four stages of corporate growth that I had identified, the itinerant attractor seemed to explain why, at some point, signals from inside and outside the company induce behavior patterns of the past, that is, not just those that represent the right response but, unfortunately also, those that are counter productive. Thus, the "itinerant attractor" appeared to represent the chaotic principle behind corporate culture. All in all, the alley of complex systems led to the following insights:

i.   Orderly behavior determines *identity*.
ii.   Orderly behavior spreads through *chaos*.
iii. Orderly behavior hinges on *simultaneity*.
iv.   Orderly behavior follows *natural selection*.
v.   Orderly behavior generates *memory*.

The findings of the Gang of Five provided new depth to the stages of organizational development that I had identified. They showed which principles of "chaos theory" prevail in the four stages of orderly behavior development, e.g. Benzi, Sutura and Vulpiani contributing the notion of "stochastic resonance" in the (first) stage of uncertain growth.[96] They also showed that complex systems typically alternate between desynchronized and synchronized states.[97] So, as populations of synchronized behavior patterns rise and decline, the growth of simultaneity predictably traces the bell-shaped growth-rate curve.

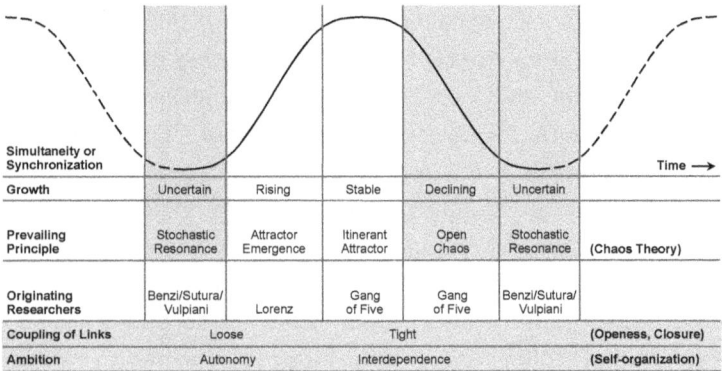

Simultaneity or Synchronization					Time →
Growth	Uncertain	Rising	Stable	Declining	Uncertain	
Prevailing Principle	Stochastic Resonance	Attractor Emergence	Itinerant Attractor	Open Chaos	Stochastic Resonance	(Chaos Theory)
Originating Researchers	Benzi/Sutura/ Vulpiani	Lorenz	Gang of Five	Gang of Five	Benzi/Sutura/ Vulpiani	
Coupling of Links	Loose		Tight			(Openess, Closure)
Ambition	Autonomy		Interdependence			(Self-organization)

Figure 5 The emergence of orderly behavior follows the same 4 stages.

While corporate organizations inspired the stages that I had identified, complex systems inspired those of the Gang of Five. This meant that these stages were much more common than I had expected. Nested at all levels, they describe the development of orderly behavior across many fields. So, as "fractal" of chaos-supported orderly behavior emergence, the signature of four stages unfurls wherever and whenever orderly behavior develops. Human organization, therefore, is not less natural or more exclusive than other forms of orderly behavior. Although different in the extent to which labor and tasks are divided, the emergence of human orderly behavior depends on the same principles that help molecules shape convection cells, fireflies flash in harmony, and birds fly in formation. Then again, if orderly behavior in human organizations arises by itself then what is the role of leaders? As I'll discuss further on, the answer to this question eventually dawned on me.

The findings in the alley of complex systems resonated with the premise of ideas of the thinkers that I discussed earlier. They amplified Bateson's "pattern that connects", his reference to the "dance of interacting parts", and his mention of "music, repetitive with modulation". They also confirmed Varela's call, "Behaviors rather than components shape our world." I felt that I was closing in on an answer to my guiding question, "Why, how does it happen?" Yet, some fundamental insight appeared to be missing still. The study of complex systems offered but more attributes of the process that shapes our world.

The Alley Of Time

In the years that followed the publication of my paper *Temporal Leadership*, I monitored the companies that I had reported on.[98] As soon as news about these companies was published, I would check whether the events covered were consistent with the succession of corporate growth stages that I had identified. As time passed, my confidence in the predictive capacity of the four stages of corporate development grew. In speeches and *A New Leadership Ethos*, I expanded my analysis to include other forms of human organization, such as nation states and religions.[99] The predictive quality of the stages of corporate growth held up even when examining the share-price movements of companies over a period of twenty to thirty years.[100] Yet, at some early point, I felt that I was wielding the sequence of four stages like a hammer in search of a nail. Whereas these four stages predicted the development of human forms of organization, they failed to explain their lifespan. Some companies would exist for several centuries.[101] Others would vanish in a matter of years. On average their lifespan appeared to be just over a decade.[102] So, I asked myself again, what basic principle was underlying the four stages of corporate development and what explained these greatly varying lifespans?

It so happened that, years before, I had presented my book, *Resonant Corporations*, to the press at the same session where Arie de Geus presented the Dutch version of *The Living Company*.[103] In his book, de Geus, a retired executive of Royal Dutch Shell, explained that the longevity of Shell and other companies rested on their sensitivity to the environment, their cohesiveness and identity, their open-mindedness or tolerance, and their conservative financing. Later, I became doubtful of this recipe when I learned about the experiments by Prigogine. They showed that a form of organization sustains itself as long as the inequality or niche that produced it in the first place persists. Hence, I checked which companies had typically outlived others. An examination of the ten largest companies in six countries with an average age of 135 years showed that the biggest group, by far, consisted of banks and financial services companies. [104] The second and third biggest groups involved energy companies and telephone carriers. So, as anticipated, the longevity of these companies appeared to hinge on the historical niches in which these companies had settled by chance, such as virtual money-transaction monopolies, oil-field monopolies, and national grids, niches that were walled (and defended) by high-cost entry and regulatory thresholds. The behavioral traits of longstanding companies that de Geus had identified, if really true, must have followed rather than produced longevity. The controversy about the reported oil reserves at Shell[105] and British Petroleum a few years later suggested that these conditions for longevity plainly were idealizations. Moreover, the second worldwide financial crisis that followed in the first decade of the third millennium showed the selfish and plainly irresponsible conduct of financial services companies and their leaders. This conduct violated the serene rules for longevity that de Geus had identified. [106] On the other hand, these developments also illustrated that both the oil and financial services industry were destined to traverse a stage of declining growth in which the main role of leaders was to confront established organizational practices and bring back some form of operational sanity.

In spite of the greatly varying lifespan of companies, the idea that time was of the essence got hold of me somehow. I was particularly inspired by an anecdotal story from Prigogine in his book *The End of Certainty* about a speech that he had made at a commemoration colloquium in Moscow.[107] Out of the blue, he had posed what he himself considered a daring statement: "time precedes existence". Of course, Prigogine was known for having shown that time moves forward. For example, you cannot bring back a lump of sugar that has dissolved in a cup of tea because the energy dissipated or wasted when growing and dissolving a "sugar-lump organization" cannot be recouped. Another "sugar-lump lifecycle" is needed. However, Prigogine's assertion, "time precedes existence", seemed to suggest that time plays a crucial role in the emergence of existence. This did not fit the views of time, in those days. The British theoretical physicist, Julian Barbour, who didn't believe in time anyway, wrote:[108]

[We should] get away from the idea that time is something. All that exists are things that change. The behavior of rods and clocks and with it a theory of duration never emerged organically. It was simply postulated.

In an essay, the Americans, Stuart Kauffman (theoretical biologist) and Lee Smolin (theoretical physicist), appeared to be on the same line of reasoning.[109] To paraphrase Kauffman and Smolin, there is nothing in time that describes the evolution of a thing. When describing a thing that is traveling through space, time plays a role in the calculation of its speed only. Kauffman and Smolin added two more observations. It is hard to imagine how one would measure the evolution of the universe because it would require someone outside it with a clock. A similar problem arises at levels where quantum mechanics describes the chance-driven evolution of some tiny universe of particles of which the internal dynamics are not transparent either. At those levels, "emergent time" is known to be missing too.[110] At the end of their essay, consistent with the ideas of "self-organization" and "complex systems", Kauffman and Smolin concluded that time, as we know it, plays a role at a local level only. [111] In other words, our world emerges from the behavior and timing of its components. Nonetheless, they failed to explain *which* role time plays at a local level (its global role, in other words). Hence, the statement, "time preceded existence" remained a mystery.

Upon further scrutiny, I discovered that Prigogine's autobiographical notes,[112] warmly referred to a speech by the French philosopher and Nobel laureate, Henri Bergson, whose ideas had influenced the thinking of Prigogine at an early stage of his academic career. Bergson was more explicit about the fundamental role of time or duration:[113]

> *The more deeply we study the nature of time, the better we understand that duration means invention, creation of forms, and continuous elaboration of the absolute new.*

Initially, Bergson's view of duration was confusing. It painted a picture of time that was significantly different from the everyday version of time used by physicists to calculate the speed of an object. In due course, I learned that Bergson had made a distinction between a "quantitative time" that is used to calculate the speed of an object and a "qualitative time" that explains the "creation of forms":

Science has to eliminate duration from time and mobility from motion before it can deal with them.

However, by introducing a fundamental version of time, Bergson added another "dimension" to the issue. So, intrigued by the role of time in the "creation of forms" (of organization), I looked into the ideas of another disciple of Bergson, the British mathematician and philosopher, Alfred North Whitehead. In his book *The Concept of Nature*, Whitehead took the idea of time or duration a step further:[114]

Duration is a concrete slab of nature limited by simultaneity. It is the passage of nature. Simultaneity is the property of a group of natural elements, which in a sense are components of duration. [Then again,] serial time is the result of an intellectual process of abstraction. It is evidently not the very passage of nature itself.

This meant that Whitehead also identified two aspects of time, that is, one fundamental aspect, the "passage of nature", which involved "simultaneity" (and the "creation of forms"), and one everyday aspect, which relied on an "intellectual process of abstraction", a description that resembled Barbour's view. Yet, Whitehead's view triggered two new questions. What is this "concrete slab of nature" that involves simultaneity? And, which "natural elements" was he referring to?

By that time I had come to read an essay by Ichiro Tsuda about complex systems, in which he discussed the emergence of orderly neuron behavior. To me, at first, Tsuda's essay was as complex as the systems that it described. [115] So, some time after discovering his work on the Internet, I called Tsuda to ask for his clarification and he was kind enough to take my call. When, during the conversation, Tsuda referred to the orderly behavior of a cluster of neurons, I must have said something like, "You mean that these neurons are firing electrical pulses at the same speed?" "No, that is not what I meant," he said. "I meant that neurons fire all at the same time." "Aha, simultaneously?" I said. "Yes!' he said, "Orderly behavior means that neurons fire the same sequence of pulses simultaneously." He continued, "Speed, on the other hand, involves time and distance and those are not part of the definition of orderly behavior. Only, simultaneity is!" At that moment, the penny dropped with me.

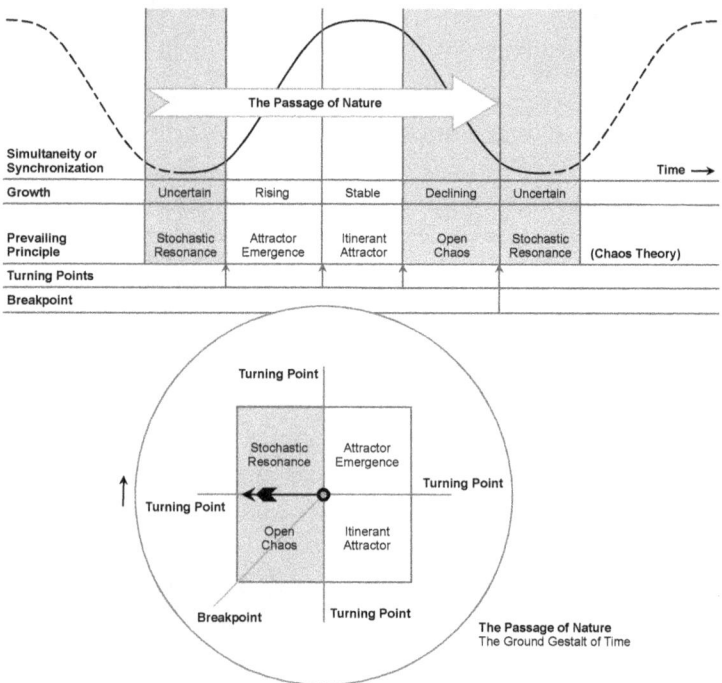

Figure 6 The clock of the passage of nature is wound by inequality.

The series of four stages, that describe the development of orderly behavior and, thus, "simultaneity", is *the* "concrete slab of nature". [116] This true fractal of chaos-supported orderly behavior emergence, which is nested at all levels and visible across many fields, *is* the "passage of nature". Since the ideas of Bergson and Whitehead were about two different aspects, if not dimensions of time, it did not take long before I realized that these stages also represented the quarters of natural duration, quarters on a clock that shows the "passage of nature" and the "creation of forms". Only an inequality of some kind would wind this clock and make it tick. A clock that displays the "passage of nature" necessarily represents the "global time" that Kauffman and Smolin were looking for. Neither contraption nor human invention but a wholly natural yardstick of emergence, it tells us what time it is from any point of view.[117] It also addresses another time-related issue identified by Whitehead:

We have to make up our minds whether time is to be found in nature or whether nature is to be found in time.

Considering Prigogine's daring statement "time precedes existence", the latter clearly applies. Palpably, nature is to be found in "the passage of nature" and, thus, in time. Every form of organization and, thus, nature is a reflection of this passage. As Whitehead hinted, nature does not contain processes and, I dare say, neither is it *a* process. *Nature is process, it's time unfurling.* Or, as Whitehead summed it up,

Spatial order is derivative from temporal order.

The above deals with the question about the meaning of the "concrete slab of nature". But, what did Whitehead mean by "natural elements"? Whitehead, a contemporary of Einstein, was somewhat annoyed by the tendency of physicists to reduce reality to "abstractions" of objects, such as "points of space and time". These abstractions, he said, were viewed as more genuine than the relations that produced them. For example, the reality of a magnetic field was derived from such point-like measures rather than from the field of relations throughout the universe that induced it.[118] This represented "misplaced concreteness". Rejecting points of space and time as legitimate abstractions of reality, Whitehead argued for "events" as abstractions of reality instead. In other words, "events" were the "natural elements" of which nature was made up. In physics, however, an event is defined as a point in (the union of) space and time, the latter referring to the time that physicists use to calculate the speed of a thing. So, the choice of "events" as "natural elements" returned the discussion to where it began: "time". Then, what is *time* (I mean the time that physicists use in their calculations)?

Barbour argued that such time does not exist. He referred to the Austrian physicist, Ernst Mach, who had said that events arise from the combined effect of all the dynamically significant masses in the universe.[119] Because of this, it is not in our power to backtrack events to their ultimate cause. So, we invented time to measure *change*. To prove his point, Barbour explained how our world might function without it. An experienced theoretical physicist, Barbour stripped the mathematical equation that describes the behavior of a tiny universe of elementary particles from its dynamic arguments, such as the center of mass of this universe, its orientation and duration (or time).[120] The resulting equation just described all the possible geometric shapes in this universe. He then explained that this universe (and, thus, our world) essentially unfolds from one such shape to the next, each shape representing a "Now". The change from one shape (or one "Now") to the next shape (or next "Now") would be inspired by how well the next shape fits the current one. As he noted rather poetically, "*Two subsequent Nows are like two lovers that seek the closest possible embrace.*" As a result, our sense of time would emerge from a phenomenon of change that hinges on the succession of best-fitting geometric shapes in the universe of all possible shapes. As Barbour admitted, when you strip an equation from its dynamic arguments, you are left with something in which energy and time do not exist. So, yes, you would have a universe of multiple shapes or "forms" but not a universe that revolves around Bergson's "creation of forms". Essentially for this reason, the path of time paved by Barbour, although highly imaginative, appeared not to lead to a viable explanation. The idea lacked causality, meaning that it failed to explain what would make Barbour's static world of shapes move forward. With reference to the earlier discussions, Barbour's world of shapes simply did not have the inequalities needed to wind the clock of the passage of nature. Moreover, along the lines of Whitehead's critique of physicists, Barbour reduced reality to a mathematical playground that includes *all* possible shapes. And, as his countryman, the theoretical physicist, Roger Penrose, apparently noted, models with multiple solutions are extremely hard to accept.[121] In all, the indiscriminate collection of shapes or forms in Barbour's world showed that a model without time lacks a provision that allows nature to exercise its energy-inspired capacity of natural selection.[122]

Then, what is the time that appears on our clocks? Before the earliest mechanical clocks were invented, people optimized their journeys by counting the days travelled. Journeys by different paths and means could be compared this way. The Earth's rotations (1 rotation representing 1 day) functioned like an imaginary yardstick: the longer the journey, the more Earth rotations would unroll. The average speed during a journey was calculated by dividing the distance travelled by the number of times the Earth had spun in the meantime. The actual distance travelled by the Earth when rotating around its axis played no role in this calculation. Similarly, the lifespan of a company in years represents the number of times that the Earth has travelled its trajectory around the Sun during the existence of the company in question. Again, the actual distance travelled by the Earth is irrelevant. What counts is the number of times that the Earth repeats this journey. So, when we chose to express the duration of events in the number of times some odd unrelated phenomenon would steadfastly repeat a journey in the mean time, we effectively invented time. Hence, our sense of time did not arise so much from change as from *repetition*. By now, we have identified phenomena that repeatedly travel much smaller distances. In 1656, based on the findings of the Italian physicist, Galileo Galilei, the Dutch scientist, Christiaan Huygens, designed a clock that relied on the steadfast swing of a pendulum.[123] At the time, the swing of the "Royal clock" with a pendulum length of just about one meter took one second. Today's atomic clocks involve the vibration of electrons in certain atoms. One of these clocks has ticks that last two 10 million billionth of a second.[124] These more and more rapidly repeated ticks allow us to clock the lifespan of ever shorter-lived forms of existence.

As Bergson and Whitehead hinted, ordinary time and (the previously discussed) passage of nature with its four quarters of simultaneity development are two different aspects time. While our ordinary sense of time *arises* from repetition, the passage of nature *produces* repetition in the shape of the repeated behavior patterns of components (as long as the inequality that produced these patterns remains in place). The passage of nature, in other words, represents the *cause* of steadfastly repeated phenomena and ordinary time represents the *symptom* of such phenomena. Thus, with reference to the Rubin vase analogy in Part 1, the passage of nature, as cause, is the ground Gestalt of time and ordinary time, as symptom, is the figure Gestalt of time (Figure 7). The assumption that ordinary time started at zero at the beginning of the emergence of our universe appears to be plausible now. Only after steadfastly repeated behavior patterns had emerged on the path of nature, (local) clocks of ordinary time appeared.

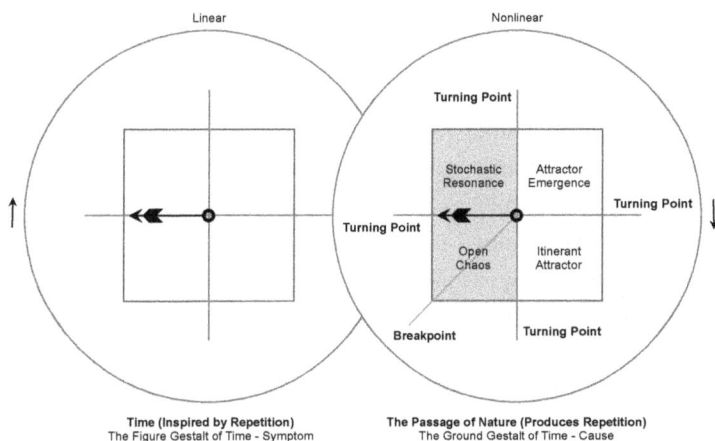

Figure 7 Two clocks showing the figure and ground Gestalt of time.

Considering the shape of the simultaneity growth curve that marks the progression of the passage of nature (Figure 6), the movement of the hand on its clock is nonlinear and thus may vary in speed. On the other hand, considering the steadfastly repeated behavior of some odd independent phenomenon that inspires ordinary time, the movement of the hand on the clock of ordinary time is linear and does not vary in speed. Astonishingly, the research by the German Egyptologist, Jan Assmann, revealed that the ancient Egyptians had come to a similar conclusion.[125] The Pyramid texts refer to two domains of time: "Time Great" ("a time of reality"), the equivalent of the passage of nature, and "Time Small" ("a time that goes by"), representing ordinary time. The ancient Egyptians described *Time Great* as the time that is used by the Gods in the domain "where new realities emerge", "a time of preexistence and post-existence, nonlinear and of cycles endless in the cosmic life". *Time Small* was depicted as "a dream that passes", a time that "governs on Earth".

Prigogine's hunch, "time precedes existence", turned out to have been prophetic. Indeed, time plays a crucial role in the emergence of existence but in a different way than anticipated. It represents a strange dichotomy that embraces both linearity and nonlinearity. As linear agent, it brings to mind Bateson's account, which referred to the symptoms of the organization of life forms as being "repetitive and rhythmical". As nonlinear agent, it evokes the ideas of the "Gang of Five", which led to the "fractal of chaos-supported orderly behavior emergence, which is nested at all levels and visible across many fields". In each case studied, it referred to patterns of behavior, patterns that shaped "things", such as ellipses, convection cells, thoughts, and human organizations. On the other hand, it failed to deal with the existence of the parts involved, such as molecules and particles. As a result, it appeared to be on a collision course with the Aristotelian and Cartesian premises of existence, which involve substances, corpuscles or parts that are divisible into other substances, corpuscles or parts.

The Grain That Caused An Avalanche

Frankly, when the penny dropped about the duality of time, I was amazed. Suddenly, the four stages of corporate growth that I had studied for so long had become a signature of the "path of nature". This meant that these four stages not only depicted the emergence of corporate reality but also of reality across many other fields. Of course, "behavior patterns" were at the heart of these phenomena of reality and "emergence" involved a process of energy conversion that induced orderly behavior and wasted energy (or produced "entropy")[126] as "transformation fee" in the process. I reported this understanding in *A New Dimension of Time*,[127] and included an analysis of the paleo-physiological and historical anchors of our worldviews.[128] I was captured by the possibility that the two aspects of time might forebode a basic shift in our thinking, a shift from an Aristotelian view (reality as a hierarchy of substances that are divisible) to a transformation-inspired view (reality as transformation, reality as becoming).[129] In short, I clustered both early and contemporary doctrines into two categories: platform-thinking doctrines (hierarchies having platforms or levels) and transformation-thinking doctrines.

I identified two paths of research in theoretical physics that dealt with matters of reality as ultimate recent examples of these categories: "Loop Quantum Gravity", as transformation-thinking doctrine, and "String theory", as platform-thinking doctrine. It was not the goal of my search (nor am I qualified) to comment on the rights and wrongs of either of these doctrines. However, in view of my quest for a perspective of reality that is rooted in behavior patterns, these examples seemed promising. Publications by the American theoretical physicist, Lee Smolin, and the Italian theoretical physicist, Carlo Rovelli, introduced me to the theory of Loop Quantum Gravity.[130] Both Smolin and Rovelli contributed some of the crucial pillars that uphold the theory. Loop Quantum Gravity describes gravity as comprising small discrete quantities or "quanta" of gravity. It is background (or platform) independent, which means that it does not require a speck of space at the onset, in which a quantum of gravity does its thing. Quite the opposite, Loop Quantum Gravity explains how gravity-field loops weave "spinfoam networks" that consist of very small "granules of space".[131] Loop Quantum Gravity, in other words, describes the spontaneous emergence of the very commodity that we so easily take for granted: *space*. As stated by Rovelli, the British theoretical physicist, Chris Isham, whom I had the privilege to meet twice in London eventually, had long pointed out the mathematics behind this process.[132] Rovelli also emphasized that "connectio-dynamics" rather than "geometro-dynamics" are central to Loop Quantum Gravity,[133] meaning that the emergence of reality arises from interactions that hinge on relations rather than on shapes or forms.[134] Different in purpose from Loop Quantum Gravity, String theory (which depends on geometro-dynamics) attempts to unify gravity with the other three basic forces of nature.[135] It depicts a tiny one-dimensional string, of which the oscillation behavior determines its role as elementary particle, each force of nature represented by some particle. Smolin and the American theoretical physicist, Peter Woit, pointed to the serious weaknesses of String theory.[136] It is background (or platform) dependent. At the onset, it assumes a speck of space, in which a string can vibrate. It also requires many more than the usual three spatial dimensions to nurture a whole hierarchy of particles, some of which have not yet been accounted for.[137] Even so, as Rovelli noted regarding the topic of background dependence, Loop Quantum Gravity and String theory may well be complementary, the former producing the background (or platform) for the latter.[138]

The comparison between Loop Quantum Gravity and String theory produced a perspective of reality that seemed to come very close to the reality produced by the sequence of the four stages of behavior-pattern emergence (the ground Gestalt of time). Even phenomena of reality as basic as the commodity of space depend on some harmony of simultaneous behavior patterns involving gravity-field loops. Thus, things do not shape behavior patterns but behavior patterns shape things. Putting it differently, connectio-dynamics rather than geometro-dynamics "write" the reality that we observe. However, as psychologists, such as Rubin, illustrated, shapes still dominate our perception of reality. So, scientist or not, we tend to behave like geometry addicts. Rubin's example of the vase and the two symmetrical faces that shape it illustrates how difficult it is to see the world beyond a shape, however close that world might be.

The research by Smolin, Rovelli and their distinguished colleagues fascinated me. At the same time, Tsuda's ideas about order being rooted in harmony and simultaneity rather than in spatial matters were roaming in my thoughts, as did the related energy-conversion process that Prigogine had identified. The temptation to seek feedback on how these ideas had led me to another aspect of time became too big eventually. So, I called Rovelli of whom I knew that he worked as professor at the Aix-Marseille University. When Rovelli answered his phone, he was on his way to London by train to meet Chris Isham, professor and theoretical physicist at Imperial College. Rovelli was kind enough to allow me an interview, knowing that I was a novice. Over lunch, in the harbor of Marseille, I explained what I had learned, having sent him a copy of my manuscript before. As I could have expected, Rovelli reacted in a subtle, even tactful way, guiding me to the work of philosophers on these matters, particularly the ancient Greek philosopher, Anaximander.[139] He also encouraged me to talk to Chris Isham whom he described as open to ideas from outside the realm of theoretical physics. Following his advice, I contacted Chris Isham some time later. Isham, also, impressed me by his generosity and willingness to hear me out and point me in new directions. As author of authoritative articles, such as *Emergence of Time in Quantum*

Gravity,[140] Isham was by no means unfamiliar with the duality and arguments of time. Inspired by the views of Whitehead and an imaginative contraption from the British artist, John Latham, he had even co-authored an article, which explores the idea of time as "time of being" *and* "time of becoming". [141]

> *There are two main uses for the concept of time in physics: first, as the parameter in temporal logic, labeling the points of "being"; and, second, as the parameter in the equation of dynamics, where it refers to the notion of "becoming". [Then again,] a key question for theoretical physics concerns the appropriate mathematical structure that is used to represent the idea of time.*

In line with the above, I recall Isham saying something like, "To judge a new idea, I need to see the mathematics first". Smolin, when working as visiting professor at Imperial College, also confirmed the possibility of a "time of becoming".[142]

> *Time may be none other than the process, which constructs not only the universe but also the space of possible universes relevant for observations made by local observers.*

The idea that time is "the process, which constructs" is consistent with the ground Gestalt of time, although Smolin's wording might suggest to an uninformed reader the mistaken idea that nature *constructs* things. Eventually, Rovelli published an article with the unambiguous title, *Forget time.* Seemingly in contrast with the latter, Rovelli stated in his conclusion: [143]

> *The peculiar properties of the time variable [in an equation] are of thermodynamic origin [that is, energy-conversion origin].*

Again, this appeared to be in line with the definition of the ground Gestalt of time. Not content with what I had found, particularly in view of Isham's pertinent remark, I searched for other literature, on the topic of "resonance". Resonance plays a crucial role in the passage of nature because it explains how behavior patterns might be amplified when they interact with one another. An article by the Russian physicist, Eugene Maslov, appeared to build on this idea.[144] Maslov explored the outcome of his mathematical model for the transition of matter, such as the transition from gas to liquid. Ignoring the minimal quantity of transformation-energy wasted in such processes (entropy) and relying on the oscillating dynamics of transitions where one state of matter slowly replaces another, Maslov neatly arrived at four "stages of evolution" that resembled the four stages of the ground Gestalt of time. Like the cycle of behavior-pattern emergence, Maslov's four stages led to the next cycle through a stage of chaos. He then suggested a parallel with transitions at subatomic levels asserting that these stages could be proven to "give rise to a stationary foam-like spatial structure" that "could manifest itself as the cellular structure of the universe observed at super scales". This remarkable claim reminded of the very thing that distinguished Loop Quantum Gravity but in a way that was more transparent and consistent with the ground Gestalt of time. I found subsequent articles by the same author dealing with cosmological issues that showed the constancy of his path of research,[145] a path that could be traced back to the ideas of the American physicist and Nobel laureate, Ken Wilson, who had inspired the gravity-field loops at the heart of Loop Quantum Gravity in the first place.[146] The crux of Maslov's research was the explanation of physical realities across fields and at multiple scales *as temporary coherent behavioral phenomena.*

At this point, I realized that I had been adding grains of insight to a pile of grains that had not really changed shape. Still thrilled to have stumbled onto the duality of time, I was restless. Something seemed to be lurking in the background, an ordinary insight rather than some esoteric phenomenon that could reveal the nub of nature. As physicist by education, I was aware of matters, such as General Relativity and the quantum world at microcosmic levels, matters that aren't exactly commonplace. But, these matters refer to the *features* of reality rather than to reality itself. All along, the remarks by Bateson and Tsuda kept my mind's eye focused. I often thought of Bateson's idea of symmetry across scales. Indeed, I felt that we shouldn't have to search for the essence of reality at micro- or macrocosmic levels. The essence of reality should be observable at our level too! Bateson's ultimate clue was also ever present in my mind:

> *The right way to begin to think about the pattern that connects is to think of it as a dance of interacting parts.*

Tsuda's observation that distance, a spatial measure, and time were not part of the definition of orderly neuron behavior had morphed into the idea that spatial features were secondary, symptoms rather than cause. Maslov's claim and Loop Quantum Gravity strengthened this idea. Of course, this meant that the essence of reality might not be about space, forms, and shapes (a premise that will no doubt upset those charmed by the views of Plato, who believed that forms precede existence).[147] For this reason, it was curious when our South African architect, Keith Skinner, whose business it was to deal in shapes and forms, handed me the grain of insight that would eventually produce the avalanche that confirmed this. Skinner, a friend, had brought me his January 2007 issue of the National Geographic, which included a colorful article about snowflake shapes.[148] He said that the article had reminded him of a conversation, in which I had referred to the relation between environment, behavior patterns, and spatial features, no doubt weary about the latter. The article reported on the research by the American physicist, Kenneth Libbrecht, who had become famous for his snowflake pictures. My eye instantly fell on a diagram that stretched across the length of the page and showed several distinct snowflake shapes alongside a vertical thermometer with a temperature scale that ran from the freezing point downward. The diagram clearly showed that snowflakes, grown in a process chamber where the conditions, such as the temperature, had been kept within a narrow range, all grew with a distinctive shape. In other words, under stable conditions, the temperature level determined the shape of a snowflake. I was surprised to read in the article that Libbrecht did not know why this was so. As discussed in Part 1, upon further scrutiny, I learned about his extensive research and understood where he was coming from. Yet, the moment that I saw the diagram, my mind translated temperature into the oscillation behavior of the water molecules that had settled.[149] Most water molecules must have had the same such behavior because the temperature had been kept stable. Hence, the relation between temperature and snowflake shape could effectively be completed with the link between oscillation behavior and snowflake shape, the former being the cause, the latter being the symptom. Putting Bateson's words to use, the "dance" of frozen water molecules that had settled was "the pattern that connects". Indeed, the congruently vibrating molecules were like virtual dance partners whose simultaneous movement "drew" the distinctive pattern of a snowflake on a three-dimensional dance floor. The causal connection, of which we only see the final state (snowflake shape), looks like this:

Temperature inequality ➜ water-molecule vibrations ➜ snowflake shape

In *A New Leadership Ethos*, in which I reevaluated the role of leaders after introducing another perspective of organization, I identified several new examples of spontaneously emerging phenomena.[150] For one, if a snowflake is an organization that involves the behavior patterns of frozen water molecules then a company is an organization that involves the behavior patterns of employees. Both are thermodynamic phenomena that convert energy more efficiently by adopting orderly behavior. While a temperature inequality (the temperature below freezing point) triggers and sustains the behavior patterns of frozen water molecules (which shape the snowflake), a supply or demand inequality triggers and sustains the behavior patterns of people (which shape an organization). The range of examples appeared to be limitless. An inequality might trigger and sustain anything from a flame, a thought and a convection cell to a company and a flock of migrating birds.[151] Always, repeated behavior patterns "drew" the shape. Hence, the transition from cause to symptom could be reduced to a simple universal relation:

Inequality ➜ behavior patterns ➜ shape

This straightforward "sequence of dependence" hinges on the stages of behavior-pattern development, that is, the ground Gestalt of time. The shapes drawn by best-reproducing behavior patterns during two stages (rising and stable growth) are what we usually observe as reality. The other two stages tend to be hidden from our view or, as my survey indicated, ignored. So, the idea that shape is the crux of reality is futile. Reproducing behavior patterns are the essence of reality and the shape that these patterns draw is but a symptom, an image on the path of nature. The insight that our world is not made up of components or shapes but of behavior patterns triggered an avalanche, in which the findings on my path of inquiry across different fields naturally joined in unison. In all, reality is:

i.   Behavioral (inequality-driven behavior patterns),
ii.   Emergent (on the path of nature),
iii.   Thermodynamic (energy-conversion driven)

Realities resemble *melting icebergs*[152] in an equilibrium ocean that are kept afloat by repeated (or reproducing) orderly behavior patterns far from equilibrium (the latter because each reality depends on an inequality of sorts). Realities are "melting" because, when behavior patterns reproduce, they dissipate or waste energy (entropy). In Part 3, I'll explore in more detail the relation between *inequality* and *behavior patterns* and how this relation yields adjacent realities. [153]

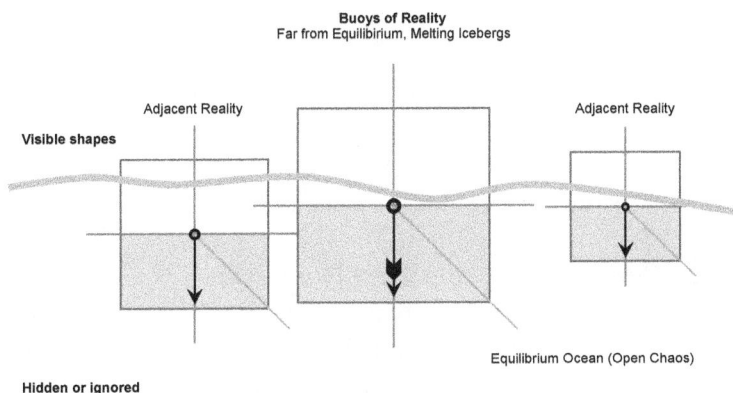

Figure 8 Realities are melting icebergs kept afloat by orderly behavior.

This new worldview means that the aged premises by Aristotle and Descartes (a world comprising substances or corpuscles) can no longer sensibly be maintained. The windows onto our world that Aristotle and Descartes had opened provide only a narrow perspective that is limited to what our senses record (Figure 9). A less egocentric perspective is needed to see existence as a wholly behavioral phenomenon. As I'll argue in Part 3, once we succeed in observing reality through the eyes of nature, we will further improve our chances. In sum, the grain that caused an avalanche was an example that simply showed shape as symptom. The avalanche surfed on waves of what were remarkably consistent views by enlightened thinkers from across a range of different fields. Last but not least, when the avalanche unfolded in my mind, I finally saw the full answer to my guiding question: "What really shapes our world?"

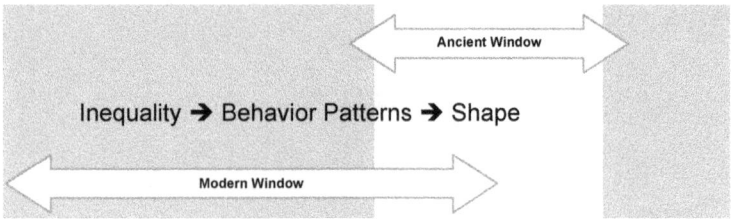

Figure 9 Two different windows onto our world.

Dots That Further Completed The Picture

Once the dust had settled after the "avalanche", I came across several new insights, which reinforced the remarkable picture of reality that emerged from the dots that I had come to connect. If the path of nature is true then it should also explain the emergence of grandiose phenomena, such as the universe. As Rovelli noted in his status update on Loop Quantum Gravity,[154] there must have been something before the Big Bang that produced the space needed to grow in the forms of matter that we are familiar with. Hence, the Big Bang was not the ultimate beginning of our universe but rather a stage in its cycle of emergence. To satisfy current cosmological observations (which show the expansion of the universe), such cycles must involve eons of collapse and inflation separated by a Big Bounce (which, looking back in time, resembles a Big Bang). However, this assumption is not without problems, one of the problems being the explanation of what has happened to the energy that was wasted or dissipated when, during the inflation of the universe, energy was converted. As the second law of thermodynamics states, when energy is converted and some of it is lost in the process as entropy, the amount of the latter will increase in the end. Consequently, some time during the cycle of the emergence of the universe, this useless form energy had accumulated to an enormous amount. Strangely enough, cosmological observations showed that, at the time of the most recent Big Bounce, the amount of entropy appeared to have been low. In *Cycles of Time*, the British theoretical physicist, Roger Penrose, explained how the Second Law might have been reset before.[155] During an eon when the universe had become rich in entropy and short of the behavior patterns needed to sustain its spatial magnitude, it collapsed. A scattered form of energy, entropy has many degrees of freedom. However, when the universe collapsed, entropy was compacted so much that it lost its degrees of freedom and, thus, the very feature that distinguished it. That is when the slate of the Second Law must have been wiped clean. Of course, as Penrose stresses, "this is a completely new phenomenon when dynamical behavior is considered." Another problem with the idea of the universe as emergent behavioral phenomenon is that, like a human organization, it needs an impulse that triggers its rise. Where did this impulse come from? Penrose argued that the "kick" that set-off the Big Bounce came in the form of a gravitational wave burst that was produced by the collapse of the universe at the end of its earlier cycle. So, the previous cycle produced the trigger that set off the next cycle. On the whole, Penrose showed the path of nature to be a plausible explanation for the emergence of the universe. What's more, as the following development shows, the idea that the Second Law can be reset coincided with another fundamental drift in physics.

In 2010, the Dutch theoretical physicist, Erik Verlinde, surprised his colleagues with an article, in which he suggested that gravity is an emergent, entropy-reliant phenomenon, rather than one of nature's presumed fundamental forces.[156] He later added that other fundamental forces might not be fundamental either.[157] Before exploring Verlinde's findings in more detail, I'll briefly reflect on the significance of his claims in the light of the state of physics today. Physicists typically describe observable phenomena in a quantitative way as accurately as possible in order to deal with these phenomena predictably. For example, to describe the trajectory of an object, they invented a measure of time as well as spatial coordinates to keep track of it. An object does not start to move by itself. So, the idea of force was introduced to explain the transition from stationary state to motion. Physicists made sure that these matters were well defined in mathematical terms and properly related to other phenomena. Eventually, the ideas of time, space, and force became embedded as indispensible to the explanation of the realities that physicists studied. However, these ideas also started to live a life of their own and became often treated as realities themselves. Einstein, for example, inspired a decades-long era in which physicists searched for a description of gravity hoping to unite it with other fundamental forces.[158] These physicists imagined the existence of minute strings and showed that their oscillation behavior would determine their role as elementary particle. One of these, they argued, would represent gravity. As mentioned earlier, physicists, such as Barbour, Isham, Rovelli, and Smolin, eventually questioned the existence of time. As a result, the idea of time has been brought to heel. Time (that is, ordinary time or the "figure Gestalt of time") is a human invention that helps observers make sense of dynamic phenomena.[159] Rovelli, Isham, and Smolin also contributed to insights, which showed that space is not a given physical reality but something that emerges at the nodes of gravity-field loops. Verlinde added that space is "a device introduced to describe the positions and movement of particles" and that gravity, as a force, can be shown to be a result of a behavioral phenomenon "using only *space-independent* concepts like energy, entropy and temperature."[160] Verlinde effectively demystified the ideas of *force* and *space* when he showed that they are the fruit of some energy-related inequality. Because the entropy increases when matter is displaced to minimize an energy-related inequality, Verlinde suggested that gravity and other forces are "entropic forces". The earlier-defined scheme for reality extended with Verlinde's items (between parentheses) now looks like this:

Inequality (➔ force) ➔ behavior patterns (➔ space) ➔ shape

In sum, the above sequence of dependence shows that both *force* and, indirectly, *space* "emerge" from some inequality. The following example of "osmosis" illustrates both Verlinde's findings and the above, extended sequence of dependence.

The legs of a U-shaped tube filled with water (Figure 10) are separated at the bottom by a membrane. Particularly found in biological systems, membranes, such as the cell walls of trees, are selective barriers that allow certain molecules to pass freely, yet block the passage of others. In this case, the membrane allows water molecules to move freely between the legs of the tube but blocks the molecules of a solvent, like salt. Salt has been added to the water in both legs but more in the left leg than in the right. So, the concentration of salt in the left leg is higher than the concentration in the right leg. Hindered by less salt molecules, the water molecules in the right leg move more freely than those in the left. In other words, the water molecules in the right leg have more "free energy". This inequality in free energy produces a flow of molecules between the legs until the free energy of the water molecules in both legs is the same. Because the membrane blocks the flow of salt molecules, water molecules flow instead.

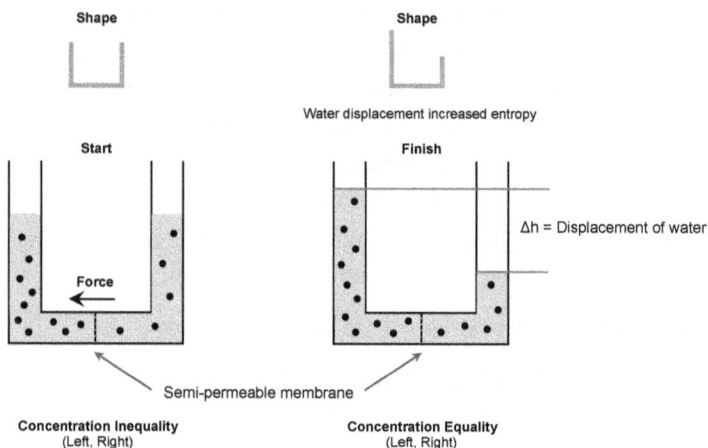

Figure 10 The potential to increase entropy induces a force.

As a result, water molecules move from the right leg to the left leg until the concentration of salt at both sides of the membrane is the same, that is, when the free-energy inequality across the membrane is zero. Back to the sequence of dependence, the force with which water molecules flowed from the right leg to the left pushed up the water level in the left leg and lowered the level in the right leg. On average, the water molecules that moved from the right to the left bump into more salt molecules. So, in the end state, the disorderly movement of water molecules or entropy is higher. This is why Verlinde said that such force is an "entropic force". The weight of the water that has been displaced represents the *force* that emerged at the start. The resulting difference between the levels in each leg (Δh) represents a spatial displacement and, thus, the *space* that emerged. This displacement also determines the specific *shape* of the connected water columns. Considering the protruding shape of the water columns in the end state, it is now obvious how trees reach for the sky in their pursuit of sunlight: They grow through osmosis! Contrary to what is sometimes assumed, "order" is not a quality of the state at the start of an experiment. Rather, it is in the orderly behavior patterns of water molecules that are sustained as long as a free-energy inequality across a membrane exists. Thus, order is a dynamic rather than a static phenomenon that governs only temporarily in situations away from equilibrium. Or, as Bateson put it poetically, order is "like music, repetitive with modulation". Regarding the force at work in osmosis, the Finish physicist, Frank Borg, showed that the size of the particles involved is not important.[161] Hence, in view of the account in this book about "the fundamental entities of which our world is composed", we can now safely reduce (or increase)[162] the size of the components in this example without changing the crux of the outcome. The question is to what extent this can be done. The sequence of dependence that describes reality again provides guidance.

Inequality (➔ force) ➔ behavior patterns (➔ space) ➔ shape

As the example of osmosis shows, space and shape arise from an inequality and the behavior patterns that it triggers. So, as symptoms, space and shape are not relevant here. Although, behavior patterns emerge as well, they are essential in that they sustain both space and shape and contribute to an increase in entropy. For this reason, the fundamental entity of which our world is composed is not determined by space or shape but by the ultimate combination of *inequality* and *behavior pattern*. As suggested in Part 1 and as I'll discuss in Part 3, one might refer to this ultimate combination as an *"inequality border dance"*. By reducing our world to a collection of inequality border dances, the apparent disconnectedness of a world of "things" can be overcome. This disconnectedness made it hard to explain a force, such as gravity, that is at work between what appear to be disconnected bodies. Consisting of a multitude of *reproducing* (visible) inequality border dances and suspended in a multitude of *brief* (invisible) inequality border dances,[163] these bodies can now communicate. This probably won't diminish the need for Verlinde's imaginative "holographic screens" that expose the emergence of gravity by storing information associated with the displacement of material bodies.[164] But, considering Verlinde's final remark, it does make our world more logical.

> *Holography is also a hypothesis, of course, and may appear just as absurd as an action at a distance [gravity].*

Discussion

There is more to our world than meets the eye. For example, there are other stages of corporate emergence than those that management gurus, leaders and consultants draw our attention to. As illustrated in Figure 8, reality is an emergent phenomenon that floats on an equilibrium ocean (a sea of societal chaos when it comes to human organizations). Without making an effort, stages below the surface may only come into view when briefly pushed up by a societal wave. Considering the unfortunate fate of the Titanic, it seems foolish to determine one's course based on what is visible above the surface believing that one's ship can withstand what lurks below. Nature unwillingly punishes such denial eventually.

Scientists, on the contrary, may well see too much in "what lurks below". As Verlinde's findings suggested, scientists may have lent too much power to phenomena, such as "gravity". In 1932, the Dutch astronomer, Jan Oort, introduced the idea of dark matter to explain the anomalies that he had observed in the motion of stellar systems.[165] Oort expected the stars visible at the outer edge of a spiral galaxy to orbit with a much lower velocity than those at the center, because that is what Newton's law of gravity predicted. But, they didn't. Assuming that Newton's law was correct,[166] Oort argued, this might indicate that the spiral galaxy was in truth much bigger than the one that he had seen. So, if the visible part of the spiral galaxy consisted of "luminous matter" then a much bigger halo of "dark matter" should make up the remainder of the spiral galaxy. Consequently, the outer edge of dark matter would orbit at the expected lower speed and the luminous visible matter at the visible outer edge (now relatively close to the center of the spiral galaxy to which the hypothetical dark matter has been added) would predictably spin nearly as fast as the stars at the center. In order to make this work, however, Oort had to assume a huge amount of dark matter, about 19 times the amount of visible matter.[167] Although there has been *no* direct evidence of dark matter whatsoever, the theory, which predicts that only 4,6% of all matter is visible, is the leading theory on the subject. Considering the earlier discussions, the theory of dark matter, which entirely hinges on *symptoms*, is a view that relies on the (hidden) shape of matter and, thus, the *figure Gestalt of reality*. This might explain why it is rife with speculation about hardly detectable particles.

Two recent insights offer explanations that appear to seek the *ground Gestalt of reality*, an explanation of our world as emergent, behavioral phenomenon, the *cause* of what we observe. One offers a *deeper* and the other a *broader* perspective. First, the earlier discussed theory of Loop Quantum

Gravity (with Loop Quantum Cosmology as branch) offers the deeper view: Space doesn't contain matter but it emerges where gravity-field loops intersect, eventually to foster different forms of matter.[168] Defined as such, space, being much more than we now assume, could explain the "behavioral anomaly" that Oort reported. Second, the Bohemian-born Australian physicist, Pavel Kroupa, and his co-workers, Marcel Pawlowski and Jan Pflamm-Altenburg, offered another view.[169] Rather than evaluating the behavior of a single spiral galaxy (the Milky Way, which hosts our planetary system), they broadened their observations to include the behavior of adjacent systems, such as satellite galaxies, globular clusters and streams of stars and gas. Kroupa showed that the interaction and collision paths of the systems in this "vast polar structure" effectively explained the distribution of velocities in the Milky Way. When explaining the celestial velocities this way, the researchers argued, there was simply no need for dark matter.

The somewhat frivolous jump from human organization to stellar systems suggests the amazing reach of the picture of reality that emerged from the connected dots. Of course, the four-quadrant system of emergence is simple, almost too simple to be true. However, this simplicity is misleading in that the four quadrants are no less fundamental. In the following part, I'll discuss in more detail how the laws of thermodynamics are at work during each of the four stages and how nature deals with these laws not having a clue of what they are. I will also show how the realities that we observe might emerge from something that is entirely behavioral. Then, in Part 4, I'll explore how the four stages of emergence are burning as a fire mark through several layers of society.

Notes

---

[1] Here is the YouTube link to the recording by Stanford University: http://www.youtube.com/watch?v=UF8uR6Z6KLc

[2] I worked for Digital Equipment Corporation (DEC) in the late eighties and early nineties. Before that, I worked eight years for Xerox on its last leg of growth just before its patented copier technology expired. Both companies experienced remarkable growth that eventually petered out.

[3] Inspired by the third stanza in Rudyard Kipling's poem "If". You'll find Kipling's poem here: http://www.kipling.org.uk/poems_if.htm

[4] Marc van der Erve, *Dynamisch Ondernemen – Strategieën voor de ontwikkeling van een flexibele organisatie*, Sijthoff, Amsterdam (1986).

[5] Marc van der Erve, *The Power of Tomorrow's Management*, Heinemann Professional Publishing, Oxford (1989).

[6] I'll explain further on why reframing often fails and what can be done to make it work.

[7] Rosabeth Moss Kanter, *The Change Masters*, Unwin Hyman, London (1989).

[8] I joined KPMG Management Consultants as partner. I did my PhD at Tilburg University (December 1993). My supervisor was Prof Dr. Jules van Dijck (who taught at the faculty of sociology and focused on matters of organization as non-executive director for various leading Dutch firms). My co-supervisor was Prof Dr. Arno de Schepper (who taught at the faculty of logistics & procurement and was managing partner with Coopers & Lijbrand, which eventually became PwC). I am eternally indebted to my academic supervisors for their guidance and support.

[9] The US PC manufacturer, Compaq, bought Digital Equipment in 1998. In 2002, Hewlett Packard bought Compaq. The labels of Digital Equipment and Compaq both disappeared from the market.

[10] Marc van der Erve, *Evolution Management – Winning in Tomorrow's Marketplace*, Butterworth-Heinemann, Oxford (1994).

[11] Following the same reasoning, achievement, to an NGO, means fulfillment of the needs of an interest group. To a nation, achievement refers to fulfillment of the needs of a people. In other words, the fulfillment of certain needs concerns a general principle that applies to all forms of human organization. Particularly, the *Austrian School* (including such economists as Ludwig von Mises and Friedrich Hayek) referred to the fulfillment of customer needs or "wants" (J. H. McCulloch, *The Austrian Theory of Marginal Use and of Ordinal Marginal Utility*, Journal of Economics, Springer Verlag (1977), Vol. 37, No. 3-4, pp. 249-280). Moreover, according to the Austrian School, the capacity to fulfill a need will

diminish progressively. In other words, as a market niche is being filled, the pulling force exerted by it reduces in strength.

[12] *Lifetime of Average S&P Company*, Business Innovation Insider (2005), No. 11.

[13] Peter Drucker, *Innovation and Entrepreneurship*, Pan Books, London (1985).

[14] James Collins, Jerry Porras, *Built To Last – Successful habits of visionary companies*, HarperCollins Publishers, New York (1994).

[15] This is an example of how established science seemed to confront my ideas. The idea that organizations are "grown to achieve" involves business cycles. Introduced by the Austrian School of economists, business cycles necessarily hinge on the *temporality of forms of organization* (the crux of my findings). Economists, such as Milton Friedman (Monetarism) and Paul Krugman (Keynesianism), both representing "established science" at some point, believed the idea of "business cycles" to be incorrect because it was not supported by evidence. Subsequent research indicated that Friedman's conclusions might have been based on "misleading data" and that Krugman's conclusion involved a "misinterpretation of the theory" (Wikipedia on the Austrian School of economics). More recent quantitative research appears to confirm the existence of business cycles: Robert Mulligan, *An Empirical Examination of Austrian Business Cycle Theory*, The Quarterly Journal of Austrian Economics (Summer 2006), Vol. 9, No. 2, p. 69-93. Mulligan ends his article with the following observation. *"The policy prescriptions of the Austrian School are unmistakable: first, never disturb the interest rate with credit expansion or monetary inflation, and second, after the first policy prescription has been violated, never interfere with entrepreneurial planners' efforts to liquidate suboptimal production plans as rapidly as possible. As long as economists and policy makers believe the business cycle can be avoided through the activism of charismatic central bankers, recessions will be inevitable."*

[16] Schumpeter | *Built to last*, The Economist (November 26th 2011).

[17] Marc van der Erve, *Resonant Corporations*, McGraw-Hill, New York (1998).

[18] Stuart Kauffman, *The Origins of Order: Self-Organization and Selection in Evolution*, Oxford University Press, New York (1993).

[19] The title *Resonant Corporations* was invented by the marketing guru and celebrated author, Philip Kotler, then professor of international marketing at Northwestern University (USA). My publisher, Hans Ritman (Scriptum in The Netherlands), invited me for diner together with Kotler who asked me about my book. Hans gave Kotler a copy of my manuscript after dinner. The next morning, Hans called me. Apparently, Kotler had suggested titling it *Resonant Corporations*. That's when I caved in and let go my title *On the Origin of Corporate Growth*, in hindsight, regrettably so.

---

[20] I had lengthy exchanges with Mark Loch, the managing partner of McKinsey & Company in South Africa, and Colin Price, practice leader in London, in 2007, after I had commented in writing on a McKinsey strategy document that Mark was so kind to share with me. The stages of declining and uncertain corporate growth were clearly not part of the equation in this strategy document. This indicated to me that McKinsey was unaware of the unavoidable sequence of corporate growth stages. In the end, after showing sincere interest initially, they somehow chose to ignore "the clock of corporate emergence" that I had presented. In a book published several years later, Price appeared to follow the same strategy as management guru, Jim Collins.

[21] Joseph Alois Schumpeter, *Capitalism, Socialism and Democracy* (1942).

[22] The longitudinal share-price evolution of companies, such as GM, GE and Apple, shows a rising share price in stages of rising and stable revenue growth and a declining share price in a stage of declining growth. In a stage of uncertain growth, the share price is undetermined, meaning that it might rise as much as it declines.

[23] Inspired by a state of stable growth, the survey participants were predictably dislodged from conditions external to their organization. A company becomes more inward looking in this stage as it organizes itself to sustain its growth path.

[24] Frank Eltman, *Evolution skepticism will soon be history*, Independent Online (29May 2012).

[25] Ilya Prigogine, *Time, Structure, and Fluctuations*, Nobel Lecture (1977).

[26] Per Bak, *How Nature Works: The Science of Self-Organized Criticality*, Copernicus, New York (1996).

[27] Roberto Benzi, *Stochastic Resonance: From Climate to Biology*, Nonlinear Processes in Geophysics, 17, (2010), pp. 431–441: www.nonlin-processes-geophys.net/17/431/2010/ doi:10.5194/npg-17-431-2010

[28] Kunihiko Kaneko, Ichiro Tsuda, *Complex Systems: Chaos and Beyond*, Springer-Verlag, Berlin (2000). I spoke by phone with Tsuda and exchanged emails with him several times. As I'll discuss later, to me Tsuda's input was the perturbation that triggered the idea of a new paradigm of reality. When studying his rather esoteric research, nearly all (not all) the pieces of the puzzle fell into place.

[29] A 5-minute YouTube video that referred to this parallel evoked the following anonymous comment: "*You can't equate temperature inequality with human supply-demand transactions for several reasons. The former is a zero-sum game. The latter is not. The latter is multivariate in how it can satisfy demand. The former is not. Humans can resist a transaction to resolve inequality. [Molecules] cannot. This creates significant problems for your meta-theory since it seems all to be founded upon equating two inequitable things.*" Of course, I was tempted to

write a reply explaining why these arguments do not really make sense if you'd give the parallel more thought. I decided not to. The comment represented "established science" and the comments section of a YouTube video is not a place to quarrel about paradigm shifts.
http://www.youtube.com/watch?v=hWAA1eoYgns

[30] Erwin Schrödinger, *What is Life? – The Physical Aspect of the Living Cell*, (1944) Based on lectures delivered under the auspices of the Dublin Institute for Advanced Studies at Trinity College, Dublin, in February 1943.
http://whatislife.stanford.edu/LoCo_files/What-is-Life.pdf

[31] I developed an App to help establish the comfort zone of leaders and the growth stage of an organization. The App also helps evaluate the leadership fit, the necessary organizational strategies, and the fit with other organizations (in case of a merger, acquisition or joint-venture). http://www.iemzine.com

[32] The executive in question was the Belgian captain of industry (and confronter-type leader), Ferdinand Chaffart.

[33] Marc van der Erve, *Temporal Leadership*, European Business Review, Vol. 16, No. 6 (2004).

[34] As I'll explain further on, human forms of organization are no less "natural" than the forms of organization that one finds in nature, such as a snowflake, a bird flock, a convection-cell organization, and a stellar system.

[35] Thomas Robert Malthus, *An Essay on the Principle of Population, as it affects The Future of Society with remarks on the speculations of Mr. Godwin, M. de Condorcet and other writers*, J. Johnson, London (1789).

[36] William Godwin, *Enquiry concerning Political Justice and Its Influence on Morals and Happiness*, Robinson, 3rd Edition (1789)

[37] Marie Jean Antoine Nicolas Caritat de Condorcet (Marquis de Condorcet), *Esquisse d'un Tableau Historique des Progrès de l'Esprit Humain*, Paris (1794)

[38] Pierre-François Verhulst, *Notice sur la loi que la population poursuit dans son accroissement*, University of Ghent (1838).

[39] http://en.wikipedia.org/wiki/Demography_of_the_United_Kingdom

[40] As I write this, the European Union seems to be facing a similar transition. At the end of a growth cycle, it needs to shake off an inhibiting legacy of established societal links (treaties) for new ideas to develop. Then, restraint (austerity) is needed to create time and space for such ideas to emerge. In the end, progress will likely follow when these ideas permeate Europe and help increase its means of subsistence.

[41] Cesare Marchetti, *Kondratiev Revisited – After One Kondratiev Cycle*, International Institute for Applied Systems Analysis, Laxenburg (1988). http://www.cesaremarchetti.org/archive/scan/MARCHETTI-037.pdf

[42] Theodore Modis, *Predictions - Society's Telltale Signature Reveals the Past and Forecasts the Future*, Simon & Schuster, New York (1992).

[43] Technology such as the seed drill helped increase the seed yield substantially. http://en.wikipedia.org/wiki/Seed_drill

[44] http://en.wikipedia.org/wiki/Diminishing_returns

[45] As I noted before, when I met representatives of McKinsey & Company in 2007, they appeared to believe that they could restore corporate success through a recipe of remedies that ignored the successive stages of corporate growth.

[46] I was shocked to learn in a recent publication by the Club of Rome that the author had based his monetary recommendations on the belief that the clock of societal emergence could simply be turned back (p. 84). Four years before, I had written to the author of the report to explain that this did not make sense. When the author responded that he didn't have time to follow up my advice, I begged him to consider the matter once more. The arguments provided in the report to maintain this belief were cosmetic. The author had changed the label of the Y-axis from *Returns* to *Sustainability.* This is a *contradiction in terms* because, considering the natural progression of the curve of diminishing returns, a maximum of returns is simply not sustainable. Second, the author had allowed the X-axis to run from *Resilience* to *Efficiency.* This again is a *contradiction in terms.* As I showed in my doctoral thesis, when society progresses from S-curve to S-curve, it swings from chaos and resilience to some form of stagnation interrupted by odd stages of order and efficiency. Indeed, as I'll explore in Part 4, the idea of sustainability is simply not sustainable. Bernard Lietaer et al, *Money and Sustainability – The Missing Link*, Triarchy Press, Axminster (2012).

[47] Further on in this book (Part 2 and 3), I will discuss the energy-conversion-related or thermodynamic reasons behind this statement.

[48] Alfred Russell Wallace, *On the Law which has Regulated the Introduction of New Species*, Annals and Magazine of Natural History (1855).

[49] Arthur Koestler, *The Act of Creation*, Arkana, London (1964, 1989), pp. 141-142.

[50] Alfred Russel Wallace, *My Life*, Chapman & Hall (1905).

[51] Charles Darwin, *On the Origin of Species - by Means of Natural Selection or the Preservation of Favored Races in the Struggle for Life,* John Murray, London (1859).

---

[52] Four years before Darwin published *On the Origin of Species*.

[53] Alfred Russel Wallace, *On the Tendency of Varieties to Depart Indefinitely From the Original Type* (1855).
http://people.wku.edu/charles.smith/wallace/S043.htm

[54] Edward O. Wilson, *The Social Conquest of Earth*, Liveright Publishing Corporation, New York (2012).

[55] Colin Woodard, *Book Review: "The Social Conquest of Earth," by Edward O. Wilson*, The Washington Post (13 April 2012).

[56] Vanessa Thorpe, *Richard Dawkins in furious row with EO Wilson over theory of evolution*, The Observer, (Sunday 24 June 2012).

[57] Paul Bloom, *The Original Colonists*, The New York Times (11 May 2012). No doubt, Bloom refers to *The Origin of Virtue* by Matt Ridley, Viking (1996). In this book, Ridley explains how society emerged from states of tit-for-tat behavior (a variant of the prisoner's dilemma) to reciprocity, trust and altruism. These states have later been explored through computer simulations extensively.

[58] Marc van der Erve, *A New Leadership Ethos – The Ability to Predict*, Antwerp (2008), pp. 112–113.

[59] Marc van der Erve, *The Mathematical Model and Simulation of a Nonlinear Real-World System*, Werktuigkundig Laboratorium voor Meet- en Regeltechniek, TU Delft (1974). This was a bachelor's degree thesis, which discussed the mathematical description and simulation of a rather complex nonlinear mechanical system.

[60] Gregory Bateson, *Steps to an Ecology of Mind*, University of Chicago Press (1972) and *Mind and Nature - A Necessary Unity*, Bantam Books, London (1979).

[61] Needless to say, Bateson's book had more to offer (for example, the notion of "deutero learning", which I referred to in another publication). However, the example quoted specifically helped me refine my research objectives.

[62] In 2007, Probst also became Managing Director and Dean of the Global Leadership Fellows Programme of the World Economic Forum. Probst is Professor of Organization & Management at the University of Geneva.

[63] Gilbert Probst, Hans Ulrich, *Self-Organization and Management of Social Systems*, Springer Verlag, Heidelberg (1984).

[64] Scott Camazine, Jean-Louis Deneubourg, Nigel Franks, James Sneyd, Guy Theraulaz, Eric Bonabeau, *Self-organization in Biological Systems*, Princeton University Press (2001).

[65] I'll evaluate in more detail the genetic perspective further on.

[66] Ichiro Tsuda, *Toward an interpretation of dynamic neural activity in terms of chaotic dynamical systems*, Behavioral and Brain Sciences, Vol. 24/5 (2001), pp. 793-847.

[67] Hermann Haken, *Synergetics - Introduction and Advanced Topics*, Springer, Berlin (2004).

[68] http://www.scholarpedia.org/article/Synergetics, http://www.springerlink.com/content/q5741m

[69] Apart from this condition, other conditions must be in place for this to happen.

[70] Ilya Prigogine et al, *Law of Nature and Time Symmetry Breaking*, Annals of the NY Academy of Sciences, 879 (1999), pp. 8-28.

[71] Francisco Varela, *Two Principles of Self-Organization*, in: Gilbert Probst, Hans Ulrich, *Self-Organization and Management of Social Systems*, Springer Verlag, Berlin (1984).

[72] Ross Ashby, *Principles of the self-organizing system*, in: *Principles of Self-Organization: Transactions of the University of Illinois Symposium*, Heinz Von Foerster et al, Pergamon Press, London, (1962) pp. 255-278.

[73] Humberto Maturana, Francisco Varela, *Autopoiesis: The Organization of the Living*, D. Reidel Publishing Company, Dordrecht (1972).

[74] Humberto Maturana, Francisco Varela, *The Tree of Knowledge – The Biological Roots of Human Understanding*, Shambhala Publications, Boston (1992).

[75] Rod Swenson, *Autocatakinetics, Yes – Autopoiesis, No: Steps Toward A Unified Theory of Evolutionary Ordering*, International Journal of General Systems, Vol. 21 (1992), pp. 207-228.

[76] Whereas Clausius looked at the grand picture, the Austrian physicist, Ludwig Boltzmann, imagined in 1877 what might happen microscopically. Boltzmann concluded that entropy is a measure of disorder or chaos (scattering, in other words). In fact, Boltzmann viewed gas molecules as colliding billiard balls in a box, noting that with each collision non-equilibrium velocity distributions (groups of molecules moving at the same speed and in the same direction, *orderly moving molecules in other words*) would become increasingly disordered leading to a final state of macroscopic uniformity and maximum microscopic disorder or the state of maximum entropy (where the macroscopic uniformity corresponds to the destruction of all field inequalities): Ludwig Boltzmann, *The second law of thermodynamics*, Populare Schriften, Essay 3, address to a formal meeting of the Imperial Academy of Science (29 May 1886).

[77] Instead of "inequality", Swenson uses the more conventional but, as I discovered, less telling term "field potential". In physics, both "field potential" and "gradient" are used to identify some kind of inequality. I started using the

term "inequality" when a McKinsey consultant failed to grasp what I meant by "gradient". The benefit of using "inequality" rather than "gradient" or "potential" is that it explains more explicitly what nurtures orderly behavior (and, thus, forms human organization) in societal cases. Paradoxically, by seeking to reduce inequality in society, we tend to destroy its ability to sustain itself through the inequality-inspired emergence of new forms of organization. Inequality sustains human organization.

[78] The empirical evidence provided by Ilya Prigogine supports Swenson's statement.

[79] Lynn Margulis, Dorion Sagan, *What is Life?* University of California Press, Berkley (1995, 2000).

[80] Lynn Sagan (later Margulis), *On the origin of mitosing cells*, Journal of Theoretical Biology, Vol. 14, Issue 3 (March 1967), pp. 225-274.

[81] In the same process, the engulfed green alga loses its flagellum and the skeleton that determines its shape and coherence.

[82] Lynn Margulis, Dorion Sagan, Marvelous microbes, Resurgence 206 (2001), pp. 10-12. Their proposition is consistent with Haken's idea of synergetics (which stands for "working together").

[83] Adrian Bejan, *Advanced Engineering Thermodynamics*, Wiley, New York (1997).

[84] "Flows" typically are natural and analogue. The term "constructal" is digital and solicits an agent that constructs (something that I took issue with in Part 1).

[85] Tom Peters, *Thriving on Chaos: Handbook for a Management Revolution*, Alfred A. Knopf, New York (1987).

[86] Kirkus Reviews: http://books.google.be/books?id=dkdkvCSR4YQC&hl=nl&sitesec=reviews

[87] Ichiro Tsuda, *Toward an interpretation of dynamic neural activity in terms of chaotic dynamical systems*, Behavioral and Brain Sciences, 24(5) (2001), pp. 793-847.

[88] On his website, Ichiro Tsuda mentions that he has an Erdős number of 4: http://cls.es.hokudai.ac.jp/~tsuda/en/index.html. Called after Paul Erdős, a prolific writer of essays on mathematics, the Erdős number is "a tongue-in-cheek measure of mathematical prominence" instituted by friends after the death of Erdős: http://en.wikipedia.org/wiki/Erdős_number. The Erdős number typically ranges from 1 to 15 and more. So, a rating of 4 (in a series of Erdős-Renyi-Szentagothai-Erdi) is pretty impressive. It is a measure of the impact of Tsuda's work.

[89] Kunihiko Kaneko, Ichiro Tsuda, *Complex Systems: Chaos and Beyond*, Springer-Verlag, Berlin Heidelberg New York (2000, English translation), pp. 108, 180. Kunihiko Kaneko, Ichiro Tsuda, *Chaotic Itinerancy*, Chaos (in: Focus Issue on Chaotic Itinerancy), 13(3)(2003), pp. 926-936.

[90] This process is called "stochastic resonance": Luca Gammaitoni et al, *Stochastic Resonance*, Reviews of Modern Physics (January 1998), Vol. 70, No. 1.

[91] Steven Strogatz, *Sync: The Emerging Science of Spontaneous Order*, Hyperion, New York (2003).

[92] Like Swenson, Tsuda refers to the minimization of local "field potentials" in his analysis of neuron cluster behavior in the brain. Ichiro Tsuda, *Hypotheses on the functional roles of chaotic transitory dynamics*, Chaos, No. 19, 015113-1 – 015113-10 (2009).

[93] Behavior patterns are not "attracted" to the most-effective pattern, locally. Rather, the most-effective pattern simply out-reproduces and, thus, survives the less-effective ones. This shows as a succession of patterns that converge to the shape of a most-effective behavior pattern (most effective in minimizing a local inequality).

[94] Ichiro Tsuda, *Hypotheses on the functional roles of chaotic transitory dynamics*, Chaos, No. 19, 015113-1 – 015113-10 (2009). These temporarily evoked memories are called "attractor ruins". When a signal triggers a journey from one attractor ruin to the next, this journey is referred to as an "itinerant attractor". Attractor ruins are "Milnor attractors", attractors that are "kicked out by tiny perturbations": Kunihiko Kaneko, *Dominance of Milnor attractors in globally coupled dynamical systems with more than 7 degrees of freedom*, Physical Review, E 66, 055201(R) (2002).

[95] Alfred North Whitehead, *The Concept of Nature* (Lectures delivered in 1919 at Trinity College, Cambridge), Cambridge University Press (1920), p. 143.

[96] Benzi, Sutura and Vulpiani introduced the term "stochastic resonance". The American mathematician and meteorologist, Edward Lorenz, identified the "attractor": http://en.wikipedia.org/wiki/Edward_Norton_Lorenz. The Gang of Five introduced the terms "itinerant attractor" and "open chaos", the latter being a "high-dimensional" form of chaos with many degrees of freedom. Open chaos can thus amplify a broad range of faint signals: Kunihiko Kaneko, Ichiro Tsuda, *Complex Systems: Chaos and Beyond*, Springer-Verlag, Berlin Heidelberg New York (2000, English translation), pp. 188-189.

[97] Kunihiko Kaneko, Ichiro Tsuda, *Complex Systems: Chaos and Beyond*, Springer-Verlag, Berlin Heidelberg New York (2000, English translation), p. 187.

[98] Marc van der Erve, *Temporal Leadership*, European Business Review, Vol. 16, No. 6 (2004). In this paper, I evaluated (and predicted) the (next) state of development of ABN AMRO, Apple, General Electric, and Marks & Spencer.

[99] Marc van der Erve, *A New Leadership Ethos – The Ability to Predict*, Antwerp (2008). In this book, I analyzed the role of leaders in the early stages of development of Buddhism, Christianity, and Islam. I added an analysis of companies in the auto industry, such as Toyota and General Motors. I also included an assessment of the roles of historical and present-day leaders in the USA, United Kingdom, South Africa, the Russian Federation, and Zimbabwe. I particularly predicted the electoral victory of the US President Barack Obama. In subsequent slideshows, I added countries, such as North Korea, and even the evolving "society" of theoretical physicists.

[100] I investigated the share-price movements of Apple, General Electric and General Motors over a period of twenty to thirty years. The share price depends on the (by-the-market) perceived corporate growth potential. The share price appeared to rise in stages of rising and stable growth and decline in a stage of declining growth. The share price was undetermined in a stage of uncertain growth.

[101] The often-quoted Finish-Swedish pulp and paper manufacturer, Stora Enso, issued its first shares in 1288 more than 700 years ago: http://en.wikipedia.org/wiki/Stora_Enso

[102] Earlier in this text, I referred to a study by Business Innovation Insider in 2005, which showed that the average lifespan of S&P Companies was 15 years. An often-quoted report by Ellen De Rooij, *A brief desk research study of the average life expectancy of companies in a number of countries*, Stratix Consulting Group, Amsterdam (August 1996), shows an average lifespan of only 12.5 years.

[103] Arie de Geus, *The Living Company*, Harvard Business School Press, Boston (1997).

[104] Brian Gongol, *Age of the World's Largest Companies* (2005): http://www.gongol.com/research/economics/companyage/ Of the 60 largest companies (selected from 6 countries across the world, 10 from each country), 58% were banks and financial services companies. 15% were oil companies, and 12% were telecommunication and information technology companies.

[105] To boost the share price of the company and the related management bonuses, the proven oil reserves reported were some 20% higher than they actually were. Through Roelof Platenkamp, a befriended top-level Shell executive who had early informed the board about these reporting discrepancies (and was subsequently punished for this and then rehabilitated), I learned more about the skewed view of reality inside Shell at the time.

[106] As consultant to ABN AMRO at (division) board level, I learned first hand about the skewed view of reality inside the bank at the time.

[107] Ilya Prigogine, Isabelle Stengers, *The End of Certainty*, The Free Press, New York (1996).

[108] Julian Barbour, *The End of Time: The Next Revolution in Physics*, Oxford University Press (1999).

[109] Stuart Kauffman, Lee Smolin, *A Problem with the Argument of Time*: http://www.edge.org/3rd_culture/smolin/smolin_p4.html

[110] Jeremy Butterfield, Chris Isham, *On the Emergence of Time in Quantum Gravity*, in: Jeremy Butterfield, *The Arguments of Time*, Oxford University Press (first edition, 1999, paperback 2006).

[111] In a reaction to the essay by Kauffman and Smolin, Hitoshi Kitada (from the University of Tokyo) and Lancelot Fletcher demonstrated from a mathematical point of view that "the absence of global time is compatible with the existence of local time, and that the problem of time, as stated by Kauffman and Smolin, is not a pseudo-problem, but an incorrectly formulated problem." Hitoshi Kitada, Lancelot Fletcher, *Comments on the Problem of Time*, (22 August 1997): http://arxiv.org/pdf/gr-qc/9708055.pdf

[112] A translation of Prigogine's autobiographical notes can be found on the Nobel Prize organization website: http://www.nobelprize.org/nobel_prizes/chemistry/laureates/1977/prigogine.html

[113] Henri Bergson, *Time and Free Will: An essay on the immediate data of consciousness*, Dover edition, Mineola, New York (2001). Original version: George Allen & Company, London (1913).

[114] Alfred North Whitehead, *The Concept of Nature* (a collection of rather poorly edited lectures), Prometheus Books, Amherst (2004). As one reviewer observed, the rather esoteric way of formulating his intricate views made Whitehead one of the most quoted but least read philosophers of his time.

[115] Ichiro Tsuda, *Toward an interpretation of dynamic neural activity in terms of chaotic dynamical systems*, Behavioral and Brain Sciences, 24(5) (2001). Not being a mathematician and, at the time, a novice when it came to complex systems, I had to read his work several times to make sense of it. I still am grateful to Ichiro for giving me a chance to deepen my insight into his remarkable research.

[116] Marc van der Erve, *A New Dimension of Time*, Antwerp (2008).

[117] To an external observer, the passage of nature is visible in the perceived reality of the phenomenon concerned, such as an organization, a snowflake, a

flock of birds, etc. The visible leg of reality involves the stages of attractor emergence and itinerant attractors (the two quadrants not shaded in Figure 6), in short, stages in which repetition or "symmetry in time" occurs most (I'll discuss the latter further on). Another crucial external indication of emergence is the "growth rate" of an emerging phenomenon. To internal observers, on the other hand, the degree of simultaneity and repetition are telltales of the stages of emergence. Internally, the sociological focus, say between openness and closure, identity and interdependence, integration and fragmentations, is a qualitative measure of the stage of emergence.

[118] Whitehead appeared to be on the same line of reasoning as the Austrian physicist, Ernst Mach. According to Mach, locally observed inertial properties of particles arise not from some independently existing absolute space but from the combined effect of all the dynamically significant masses in the universe: Julian Barbour, *The Development of A New Dimension of Time Themes in the Twentieth Century*, in: Jeremy Butterfield, *The Arguments of Time*, Oxford University Press (first edition, 1999, paperback 2006).

[119] Masses wield the force of gravity on their immediate and far-away neighbors, both small and large. It is clear that Ernst Mach also influenced Whitehead who based his arguments on a similar observation.

[120] This is the Schrödinger wave function, called after its inventor, Erwin Schrödinger, the Austrian physicist and inventor of Quantum Mechanics. Barbour summarized the wave function with the following notation: ψ (relative configuration, center of mass, orientation, time). After stripping it from its dynamic arguments (between the parentheses), it looked like this: ψ (relative configuration). The latter describes what Barbour called "configuration space", that is, an imaginary space that includes all the wave configurations possible. Julian Barbour, *The End of Time: The Next Revolution in Physics*, Oxford University Press (1999).

[121] Barbour refers to Penrose's opinion near the end of his book (reference of which is in the previous note).

[122] I'll explore nature's energy-inspired capacity of natural selection later.

[123] It is worthwhile remembering here that Huygens's introduced an *inequality* in the shape of a weight to ensure that the pendulum sustained its unwavering swing behavior.

[124] A study has shown that the cesium fountain clock is on time to within two 10 million billionths of a second: Tim Hornyak, *British atomic clock is world's most accurate*, CNet News (27 August, 2011).

[125] Jan Assmann, *Moses The Egyptian*, Harvard University Press (1998). Professor Assmann was also kind enough to send me a copy of his speech at the

University of Liège, *La notion d'éternité dans l'Egypte ancienne*, after exchanging emails with him.

[126] Of course, "entropy" means "transformation".

[127] Marc van der Erve, *A New Dimension of Time*, Antwerp (2008).

[128] *Paleo*-physiological anchors refer to *prehistoric* or *ancient* developments affecting the functioning of organs, such as the brain and gut, developments influenced particularly by spatial (e.g. environment) and time-related (e.g. daily cycle) contexts.

[129] Along the lines of thinking by Alfred North Whitehead: Alfred North Whitehead, *The Concept of Nature*, Prometheus Books, Amherst (2004).

[130] Lee Smolin, *Three Roads to Quantum Gravity*, Basic Books, New York (2001). Carlo Rovelli, Loop *Quantum Gravity*, Living Rev. Relativity, 11 (2008), 5. http://www.livingreviews.org/lrr-2008-5.

[131] Such space granules appear at a Planck scale, that is, at a millionth of a billionth of a billionth of a billionth of a meter or 10^{-35}.

[132] As Rovelli noted in his historical overview of Loop Quantum Gravity (Carlo Rovelli, Loop *Quantum Gravity*, Living Rev. Relativity, 11 (2008), 5, p. 12.): Chris J. Isham, *Topological and global aspects of quantum theory*, in: B. S. DeWitt, R. Stora (editors), *Relativity, Groups and Topology II*, Proceedings of the 40th Summer School of Theoretical Physics, NATO Advanced Study Institute, Les Houches, France (27 June- 4 August 1983), pp. 1059-1290, Amsterdam, New York (1984).

[133] Carlo Rovelli, Loop *Quantum Gravity*, Living Rev. Relativity, 11 (2008), 5, p. 11. http://www.livingreviews.org/lrr-2008-5.

[134] The latter reminded me of the progression of "best-fitting geometric shapes" envisaged by Barbour, the viability of which I questioned earlier.

[135] Brian Greene, *The Elegant Universe*, W. W. Norton & Company, New York (2003). Four forces are believed to be fundamental: gravity, electromagnetic forces, weak nuclear forces (radioactive or beta decay), and strong nuclear forces (forces that hold protons together in a nucleus). Recent research by Erik Verlinde (discussed further on) indicates the possibility that these forces are not as fundamental as they seem.

[136] Lee Smolin, *The Trouble With Physics: The Rise of String Theory, The Fall of a Science, and What Comes Next*, Houghton Mifflin Company, Boston (2007). Peter Woit, *Not Even Wrong: The Failure of String Theory*, Basic Books, New York (2006). I found the former (Smolin's book) so much more accessible, broader, and deeper for a novice reader, like me.

[137] Between 10 to 26 spatial dimensions:
http://en.wikipedia.org/wiki/String_theory

[138] Carlo Rovelli, Loop *Quantum Gravity*, Living Rev. Relativity, 11 (2008), 5, p. 10. http://www.livingreviews.org/lrr-2008-5.

[139] Rovelli eventually published a book about Anaximander: Carlo Rovelli, *Anaximandre de Milet, ou la naissance de la pensée scientifique (Anaximander of Milet or the birth of scientific thought)*, Dunod, Paris (2009) or Carlo Rovelli, *The First Scientist: Anaximander and His Legacy*, Westholme, Yardley (2011).

[140] Jeremy Butterfield, Chris Isham, *On the Emergence of Time in Quantum Gravity*, in: Jeremy Butterfield, *The Arguments of Time*, Oxford University Press (first edition, 1999, paperback 2006), pp. 111-168.

[141] Chris Isham, Konstantina Savvidou, *Time and modern physics*, pp. 6-26 in: Katinka Ridderbos (editor), *Time*, Cambridge University Press (2002).

[142] Lee Smolin, *The present moment in quantum cosmology: Challenges to the arguments for the elimination of time* (30 August, 2000).

[143] Published about a year after we met in Marseille: Carlo Rovelli, *Forget time*, Essay written for the FQXi contest on the Nature of Time (August 24, 2008). http://www.fqxi.org/data/essay-contest-files/Rovelli_Time.pdf

[144] Eugene Maslov, *Parametric Resonance As Possible Cause Of Spontaneous Transition From Meta-stable States*, Annales de la Fondation Louis de Broglie, Vol. 26 Spécial, (2001).

[145] Vladimir Koutvitsky, Eugene Maslov, *Instability of coherent states of a real scalar field* (12 October 2005). http://arxiv.org/pdf/hep-th/0510097v1.pdf, Vladimir Koutvitsky, Eugene Maslov, *Gravipulsons*, American Physical Society, Phys. Rev. D, Vol. 83, No. 12 (17 June 2011). http://arxiv.org/abs/1106.5377

[146] In addition to loops, Wilson worked on renormalization (applying solutions across multiple scales) and condensed-matter physics involving phase transitions with regions of chaos and order, topics that typically appeared in Maslov's studies of cosmological phenomena, such as solitons (self-reinforcing solitary fields or waves) and pulsons (solitons with oscillating energy densities).

[147] The ancient Greek philosopher, Plato, divided our world into an "intelligible world" (a realm of perfect forms) and a visible world (the world in which we live with reflections of these forms). The earlier discussed ideas by Julian Barbour were based on this premise: the visible world inspired by some progression of best-fitting forms in a world containing all possible forms.

[148] Michael Klesius, *The Mystery of Snowflakes*, National Geographic (January 2007).

[149] Of course, the vibration behavior of a molecule that has settled involves wobbly oscillations in three spatial dimensions.

[150] Marc van der Erve, *A New Leadership Ethos – The Ability to Predict*, Antwerp (2008).

[151] As the autumn nears and the days get shorter, certain bird species gather to fly south in a V-shaped organization. As long as birds reproduce their flapping behavior, a bird flock's V-shape is maintained. The birds sustain this organization until they have sufficiently minimized the inequality between the conditions in the departure- and destiny-region.

[152] As I write this, I realize how much the iceberg metaphor resembles Steve Jobs' qualification of the company that he had founded: "*Apple is like a ship with a hole in the bottom, leaking water.*" Conrad Quilty-Harper, *Steve Jobs at Apple: a relentless rise in graphs and charts*, The Telegraph (6 October, 2011).

[153] Not to be confused with Stuart Kauffman's "adjacent possible configurations": Stuart Kauffman, Lee Smolin, *A Problem with the Argument of Time*: http://www.edge.org/3rd_culture/smolin/smolin_p4.html

[154] Carlo Rovelli, Loop *Quantum Gravity*, Living Rev. Relativity, 11 (2008).

[155] Penrose introduced his "Conformal Cyclical Cosmology" theory in: Roger Penrose, *Cycles of Time: An Extraordinary New View of the Universe*, Alfred A. Knopf, New York (2011).

[156] Erik Verlinde, *On the Origin of Gravity and the Laws of Newton* (6 January 2010). http://arxiv.org/pdf/1001.0785

[157] Verlinde did so during a brilliant lecture at the Perimeter Institute in Canada where he kicked off with the observation: "Gravity, and the other [fundamental] forces, are reaction forces due to the fast microscopic dynamics of the underlying system." http://streamer.perimeterinstitute.ca/Flash/8197a140-f8c8-4fb3-a676-7be339825989/viewer.html

[158] Gravity, electromagnetic forces, weak nuclear forces (radioactive or beta decay), and strong nuclear forces (forces that hold protons together in a nucleus) were so far considered as fundamental forces.

[159] According to Galileo and Einstein: Depending on the circumstances of the observer in relation to the dynamics of the phenomena that he/she observes.

[160] The formula that describes this is: $T\Delta S = F\Delta x$ (T is temperature, ΔS is the entropy difference or potential, F is force, Δx the displacement distance).

[161] Frank Borg, *What is osmosis? Explanation and understanding of a physical phenomenon*, (2003). http://arxiv.org/abs/physics/0305011

[162] One might argue that by increasing the size of components to the size of heavenly bodies, one can explain gravity as a force that emerges much like the force of osmosis. Verlinde essentially hinted this, of course. The membrane in the figure below is not really needed. The mass of the component on the left is such that it won't move a lot compared to the smaller component on the right. So, in line with Newton's law of gravity, the relative difference in mass replaces the membrane here. Most importantly, the figure (below) obeys the extended sequence of dependence that describes reality.

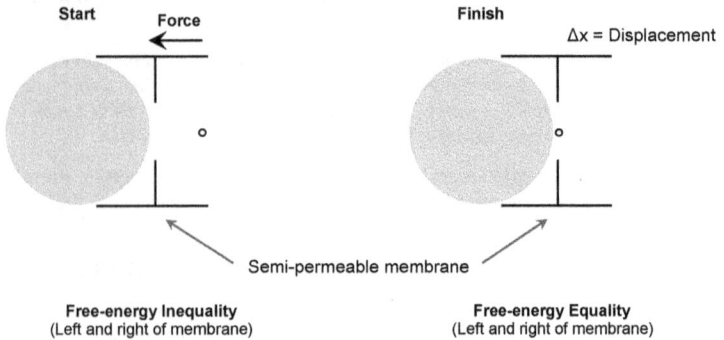

Free-energy Inequality
(Left and right of membrane)

Free-energy Equality
(Left and right of membrane)

[163] Exceptionally short-lived (invisible) inequality border dances might well come in the shape of highly energetic quantum fluctuations. Heisenberg's uncertainty principle expressed in terms of energy and time ($\Delta E \Delta t \approx h/2\pi$) shows that the shorter lived an inequality border dance, the higher the energy of its fluctuation and, thus, the greater the possibility that locally reproducing wave patterns might spontaneously emerge and become visible as particles.

[164] Entropy generates a lot of information on such screens as a result of the random displacement of the particle(s) traced. So, as Verlinde suggested, the increasing amount of information on subsequent screens is a measure of entropy and, thus, force.

[165] http://www.pnas.org/content/10/6/253

[166] Verlinde may have indirectly illustrated this in: Erik Verlinde, *On the Origin of Gravity and the Laws of Newton* (6 January 2010). http://arxiv.org/pdf/1001.0785

[167] Currently estimated distribution of matter and energy: 4,6% visible matter (the usual stuff that we are familiar with), 23% dark matter, 72% dark energy. So, some 95% of this matter and energy has never been observed directly!

[168] Matter may spontaneously emerge in absolute vacuum as a result of so-called "quantum fluctuations".

[169] Marcel Pawlowski, Jan Pflamm-Altenburg, Pavel Kroupa, *The VPOS: a vast polar structure of satellite galaxies, globular clusters and streams around the Milky Way*, Monthly Notices of the Royal Astronomical Society, 000, 1-21 (2012). http://arxiv.org/pdf/1204.5176v1.pdf

The Passage Of Nature

In my garden in South Africa, streams of ants often appear from nowhere to take advantage of some favorable environmental condition. Like us, animal species are able to spot and exploit favorable ecological niches as soon as they emerge. Considering the dots that I had come to connect, I realized that the ants in my garden follow a process that mirrors the fundamental stages of orderly behavior emergence. In stage one, in pursuit of food, ants go where no ant has gone before, many failing to return to the ant colony. This stage of deliberate chaos continues until an ant returns with a sample of food. In stage two, other ants will start following the smelly trace like Hansels and Gretels until they too find the source of food (often a dead insect). By the time these ants have returned, other ants will have left in waves along the same path, cutting corners and shortening it as much as the local conditions allow. In stage three, the traffic thickens when streams of departing and returning ants reproduce this optimal pattern of behavior over and over again. To an outsider, it seems as if this path has been etched into the collective memory of the ant colony. In stage four, when the source of food has dried up, the streams of ants reduce to trickles before they vanish as suddenly as they appeared. As the ants spread in all directions to search for new sources of food again, they return to a stage of deliberate chaos. As it happens, in my garden, as elsewhere, the end of a cycle (a dead insect) often functions as the inequality that triggers an emergent reality (orderly streams of ants), behavior patterns that reproduce long enough to be observed. At every level of nature, the same principle is at work. Particles involve reproducing patterns of energy waves that are sustained by some inequality (an environmental niche far from equilibrium). Companies involve reproducing patterns of people behavior that are sustained by some inequality (a niche far from equilibrium consisting of market and organizational conditions). In all, reality boils down to the duality of *inequality* and *behavior patterns* or, ultimately, as I'll discuss further on, to an *inequality border dance*. Because the latter is a behavioral phenomenon, so is reality. Because no dance lasts forever, reality is emergent. Because dance involves motion and, thus, energy conversion, reality is a thermodynamic phenomenon. In short, reality is:

    i.   Behavioral (inequality-driven behavior patterns),
    ii.   Emergent (on the path of nature),
    iii.   Thermodynamic (energy-conversion driven)

The Missing Principle

The realization that we are not living in a world of things but in a world of behavior patterns is a fundamental yet subtle change. It won't make our world look different in a material sense. We only know that all we see is not all that "is". Then again, the four stages of behavior-pattern emergence, once you know what they are, will become visible in most of the phenomena that we observe, no matter whether it concerns the conduct of ants, employees, neurons, molecules or waves. Once you recognize a stage of emergence, you also know the stage that comes next. As a result, our sense of prevision will improve. But, how do we reconcile a worldview, which blatantly reduces reality to some unity of inequality and behavior patterns, with the established material worldview, the subject of today's scientific theories? In *The End of Time*, the British theoretical physicist, Julian Barbour, offered a classification of theories that hinted at a link. [1] Barbour distinguished two types of scientific theories: "theories of the world" and "theories of principle". The former typically explain local physical situations and the latter refer to laws that "always hold", such as the laws that describe the conversion of energy or, in short, the laws of thermodynamics. Since thermodynamics also lies at the heart of reality as behavioral phenomenon, it allowed me to bridge the material and behavioral worldviews. In what follows, I'll explore how the laws of thermodynamics play a role on the path of nature when it generates the reproducing behavior patterns that "shape" the reality that we observe.

To show the role of the laws of thermodynamics, I'll use the earlier introduced metaphor, in which I depicted the cycle of four stages as a melting iceberg that floats in some equilibrium ocean or sea of societal chaos (Figure 11). The path that nature follows runs clockwise through the four quadrants each of which represents a stage of behavior-pattern emergence. Let's assume that we join nature on this path sometime when a local inequality arises (the bottom-right quadrant under the surface of the equilibrium ocean). A new need in the market or a drop in performance represents an inequality when it comes to human organizations. A dead insect sure is an inequality when it comes to my South African ants. On its clockwise path, nature then enters the bottom-left quadrant when behavior patterns (of people, ants, molecules, or waves) spontaneously emerge to minimize such inequality and re-establish equilibrium.

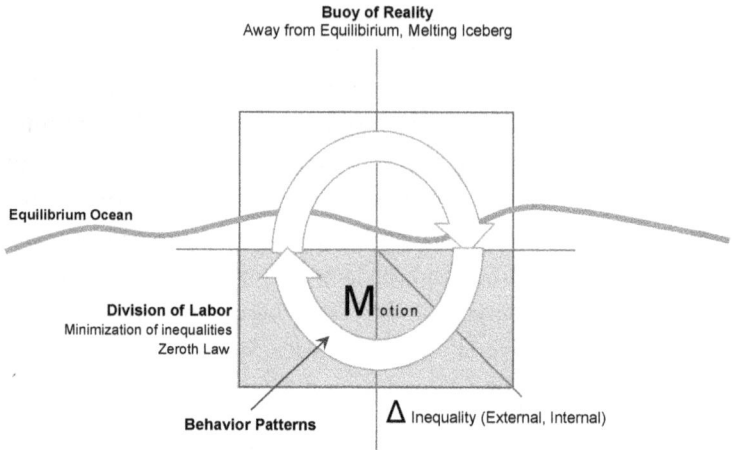

Figure 11 The first stage of behavior-pattern emergence.

Nature's propensity to minimize a local inequality and reestablish equilibrium by motion is predicted by the zeroth law of thermodynamics or, in short, the Zeroth Law. From an organizational viewpoint, the emergence of behavior patterns, albeit haphazard at first, also signals a division of labor in the reestablishment of equilibrium.[2] If an inequality is big enough and persists, equilibrium will not be achieved. At that time, nature enters the next stage of behavior-pattern emergence and moves from the bottom-left to the top-left quadrant (Figure 12). The latter rises above the surface of the equilibrium ocean. So, as nature moves to this stage of behavior-pattern emergence, it enters the realm of non-equilibrium thermodynamics where reproducing orderly behavior patterns govern in reaction to a persistent inequality (thus, in a state away from equilibrium). Most importantly, in this stage, the haphazard behavior patterns from the previous stage are thinning out when certain behavior patterns start out-reproducing others.

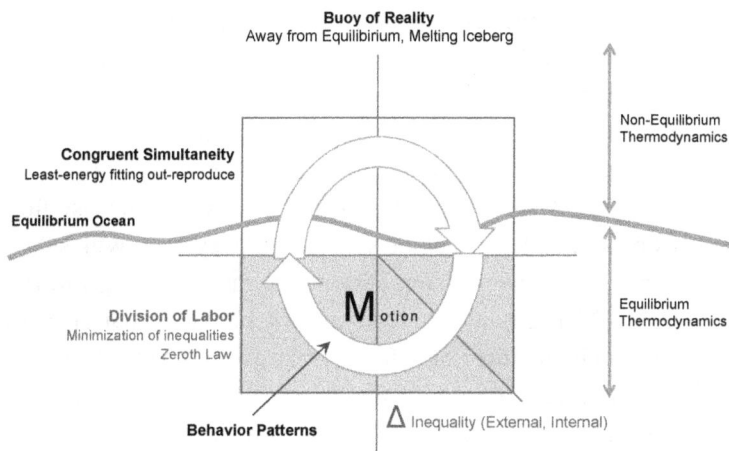

Figure 12 The second stage of behavior-pattern emergence.

No distinct law has yet been identified that describes the principle behind the natural selection of behavior patterns in this stage. It is not clear, in other words, which trait allows behavior patterns to out-reproduce others. Various findings have been tossed into the arena of competing explanations. The Norwegian-born, American physical chemist and Nobel laureate (1968), Lars Onsager, studied the effect of inequalities and discovered reversible relations between flows and (inequality-driven) forces, generally referred to as "Onsager reciprocal relations".[3] For example, a voltage inequality causes a heat flow while a temperature inequality causes an electrical current.[4] Although Onsager's research showed that inequalities produce behavior patterns, such as heat flows and currents, it didn't go as far as suggesting why certain behavior patterns are favored over others. The findings of Prigogine were more helpful in this respect. Prigogine showed that nature favors a state of least resistance.[5] So, out-reproducing behavior patterns (behavior patterns that survive) might resemble rivers in that they follow paths of least resistance. While this still doesn't reveal the trait that explains why one behavior pattern is favored over another, it does suggest that this trait involves some measure of energy efficiency. The American evolutionary and systems theorist, Rod Swenson, added to Prigogine's finding, "a system will select paths that minimize an inequality at the fastest rate given the constraints."[6] Then, which trait allows behavior patterns to minimize an inequality at the fastest rate given the constraints?

Tackling the issue from another angle, one might examine what *should* happen. In other words, which event makes the clock move forward in this stage and how does this event turn haphazard behavior patterns into orderly ones? Once more, Barbour offered a hunch that led to an answer, possibly even a law that "always holds". As discussed before, in Barbour's imaginary geometric world of all possible shapes and forms, the clock moves forward from one "Now" to the next "Now" when the shape or form of the next "Now" best matches the shape or form of the current "Now".[7] Regarding the question at hand, Barbour thus introduced "best-matching" as the trait behind the evolution of shape. Despite the progression of best-matching conditions, Barbour's model of reality failed to explain how a world of all possible shapes and forms moves forward because energy-related matters had been taken out of the equation. What is more, Barbour's model is not just "geometro-dynamic" but also "background dependent" because it assumes that forms and shapes and, thus, space, are there at the onset. So, to make the best-matching idea work in a world governed by behavior patterns rather than shapes (shapes emerge from reproducing behavior patterns), it needs to be translated from "geometro-dynamic" language into "connectio-dynamic" language.

The American physicist and snow expert, Ken Libbrecht, hinted an approach to this translation when he showed that the shape of snowflakes grown in a controlled environment hinges on temperature. The crux of the matter lies in how temperature determines the shape of a snowflake. One might expect that, in a controlled environment with a certain temperature, freezing water molecules all have the same motion energy. When water molecules no longer rotate thus freeze, their vibration behavior is a prime component of this energy.[8] So, in an environment with a certain temperature, frozen water molecules would not only share the same energy but also the same vibration behavior. The temperature, in other words, *syncs* the vibration behavior of frozen water molecules. Ultimately, the vibration behavior of individual frozen water molecules might determine the shape of the snowflake collective in much the same way as the flashing behavior of individual fireflies determines the shape of the flashing collective. Or, as Bateson told his students:

Think of it as a dance of interacting parts!

Of course, the shape of a vibration pattern or dance figure only perceptually exists. We perceive a shape or figure because the vibration pattern or dance of the parts or partners involved is reproduced over and over again, faster than our senses can distinguish. Because shape does not really exist at the level of the interacting parts or dance partners, it is not possible to distinguish whether one shape best-matches another. So, "best-matching" as criterion or trait is useless at that level. By the way, nature is not even equipped to examine the parallels between shapes at that level. Who will do the examining? Rather, as Prigogine illustrated, nature favors a state of least resistance. Considering the irritation and fatigue of dance partners that do not follow each other well enough, it is palpable that nature favors *least-energy fitting* dance patterns. So, the trait that allows behavior patterns to out-reproduce others in this stage is whether they are the *least-energy fitting* given the *local* dynamic conditions, the latter involving other behavior patterns. Least-energy fitting behavior patterns of necessity also reduce an inequality at the fastest rate given the constraints. In sum, the *Least-Energy Fitting principle* or *LEF principle* lies at the heart of the *natural selection of behavior patterns* and *non-equilibrium thermodynamics*. Shape simply is a perceptional topic that does not play a role in the passage of nature. But, when are behavior patterns least-energy fitting?

As discussed in Part 2, the clock that depicts the four stages of behavior-pattern emergence describes the rise and decline of simultaneity and orderly movement. With reference to the example of osmosis (Figure 10), the orderly movement of water molecules involved a flow through a membrane from the right leg to the left leg of a U-shaped tube. In other words, the least-energy fitting behavior pattern involved water molecules that moved both congruently (harmoniously from a direction point of view) and simultaneously. The least-energy fitting trait thus rests on *congruent simultaneity*. The latter is about parts or players that move in the same direction all at the same time in the slipstream of others and inspired by a shared inequality.[9] Hence, migrating birds, inspired by a falling temperature and shortening daytime duration, predictably all move in the same direction. They achieve the least-energy fitting behavior pattern by flying in the slipstream of the bird ahead. Players in human organizations also move in the same direction at the same time inspired by shared market and organizational conditions. They achieve the least-energy fitting behavior pattern by working in the slipstream of leaders and peers. In a snowflake, frozen water molecules vibrate congruently inspired by a common settling temperature. When water molecules settle on a snowflake, they may briefly unfreeze water molecules at the surface and jockey for the least-energy fitting dance pattern given the temperature conditions locally.[10]

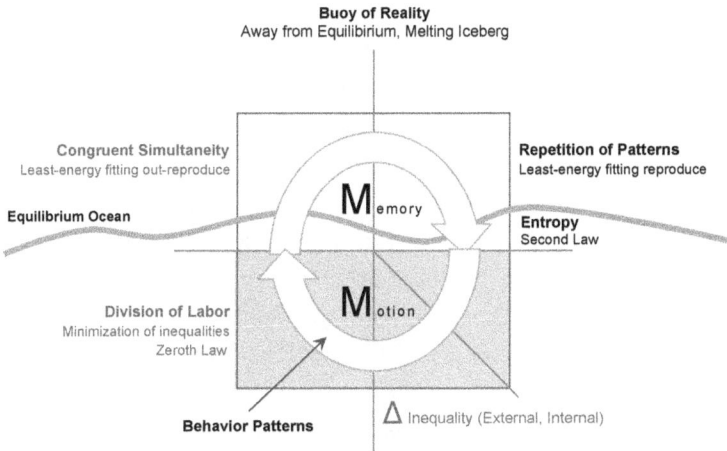

Figure 13 The third stage of behavior-pattern emergence.

On the whole, in this stage of emergence, least-energy fitting behavior patterns emerge from haphazard ones by a process of natural selection that rests on *congruent simultaneity locally*. Inspired by a common inequality and intertwined in the dance of the collective, interacting parts or players shape the whole in the eyes of an external observer.

In the third stage of behavior-pattern emergence, the least-energy fitting behavior patterns, that out-reproduced or survived less-fitting behavior patterns in the second stage, keep on reproducing as long as the inequality that kicked off the first stage endures. Reproducing least-energy fitting patterns, reliant as they are on an energy inequality, belong to the realm of "non-equilibrium thermodynamics". As energy is converted from one form into another (reproducing behavior patterns resemble the repeated movement of the piston in a steam engine), energy is wasted in the shape of entropy, the cost of transformation. A scattered, impotent form of energy, entropy belongs to the realm of "equilibrium thermodynamics" under the surface of the equilibrium ocean. The Second Law (of thermodynamics) neatly predicts its ultimate increase. Yet, the third stage of behavior-pattern emergence distinguishes itself in another way. It unveils *memory* as another core quality of nature (next to motion). A least-energy fitting, reproducing behavior pattern concerns the repetition of inequality-inspired motion. It thus is an image of the conditions that produced it. This image persists provided that the inequality that triggered it has not been minimized. As the research of Tsuda and the Gang of Five showed, our memory works just the same. A wholly behavioral phenomenon too, our memory depends on neurons that reproduce unique behavior patterns, distinctive sequences of electrical pulses that are fired in unison. [11] Each recall involves such a sequence of electrical pulses and unfolds in keeping with the four stages of behavior-pattern emergence. In all, as the progression in this stage illustrates, memory, like motion, lies at the heart of nature. Memory is not a prerogative of complex life forms nor is it an ability that magically arises from complexity. As a matter of fact, when it comes to the passage of nature, the question is not how memory arises from complexity but how complexity arises from memory.

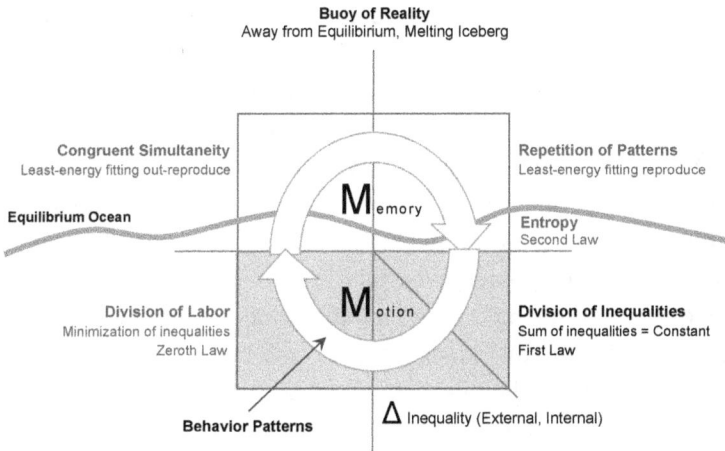

Buoy of Reality
Away from Equilibirium, Melting Iceberg

Congruent Simultaneity
Least-energy fitting out-reproduce

Repetition of Patterns
Least-energy fitting reproduce

Equilibrium Ocean

Entropy
Second Law

Memory

Motion

Division of Labor
Minimization of inequalities
Zeroth Law

Division of Inequalities
Sum of inequalities = Constant
First Law

Behavior Patterns

Δ Inequality (External, Internal)

Figure 14 The fourth stage of behavior-pattern emergence.

In *An Inquiry into the Nature and Causes of the Wealth of Nations*, the Scottish economist, Adam Smith, explored the "division of labor" and how this might improve the productive powers of a company.[12] Smith argued that the division of labor helps an organization increase the agility of its workers. If each worker takes on one sub-task, they all become more specialized in what they do. Because workers no longer need to adjust from one sub-task to another, time is saved. The division of labor also stimulates the invention of methods and tools that facilitate a subtask. However, when dividing a manufacturing task into subtasks, for example, quality-related inequalities, flow-related inequalities, resource-related inequalities, and control-related inequalities may emerge *in-between* these subtasks. Hence, to accomplish the subtask-enabled efficiencies, additional subtasks are needed, such as, management, material planning, transport and quality assurance. Oblivious of all these matters, nature pursues efficiency simply by favoring the least-energy fitting behavior patterns. Yet, these behavior patterns are not unlike subtasks in the minimization of an inequality. New inequalities emerge in-between these behavior patterns that invite their own minimization too. Hence, in the fourth stage, the cycle of behavior-pattern emergence restarts where it ends with the emergence of internal (as opposed to external) inequalities that appear in-between reproducing behavior patterns. The cycles of behavior-pattern emergence that are set off by these *internal* inequalities are what I called *adjacent realities* earlier. As Smith summed it up, the emergence of sub-tasks "is as much a consequence of the division of labor as its cause." So, the division of labor minimizes inequality by creating it. As a result, the inequality-producing process will go on forever. This is also what the first law of thermodynamics essentially predicts. It states that the total amount of energy does not change. When energy is converted at the cost of entropy, only the mix of energy forms is changing. Similarly, the sum of inequalities will not change either. Only the mix of inequality will change. When inequality is minimized, it will be replaced by inequalities that develop in-between the reproducing behavior patterns that emerge. The division of inequality therefore mirrors the division of labor. The relation between the two is one of tit for tat or reciprocity. However, there is a catch to this. Whereas the total inequality remains unchanged, the mix of inequalities will become more and more fragmented over time. Evermore delicate behavior patterns will minimize ever-smaller inequalities. To an external observer, our world thus becomes evermore complex. The American scientist, Robert Hazen, illustrated that even something as solid as a rock has evolved to become more and more complex this way. A snowflake-like phenomenon, a rock involves unique lattices of vibrating molecules called minerals. Hazen identified that since the Earth was formed 4,5 billion years ago some 4000 unique minerals have emerged. Both the number and complexity of these crystalline structures are likely to grow further.[13]

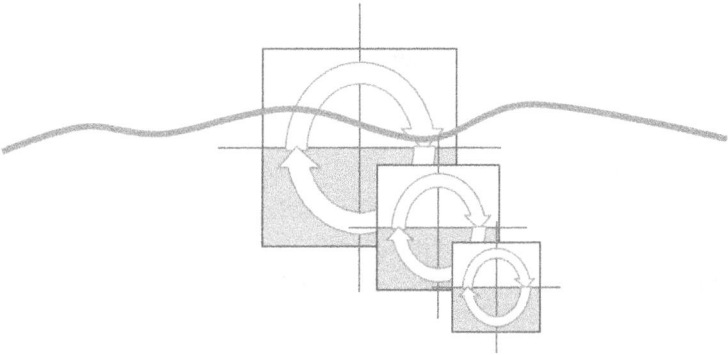

Figure 15 The ground Gestalt of time is a time fractal.

Fractal Of Time

It is not surprising that in a society dominated by a geometro-dynamic worldview, "fractals" are used to illustrate the logic behind geometry. A fractal typically is a simple shape that re-appears at smaller and smaller scales. Equally, fractals can be assembled using some self-repeating procedure to produce the complex shapes of larger objects, such as trees, snowflakes and coastlines. The Polish-born French American mathematician, Benoît Mandelbrot, coined the term, fractal, to describe the self-similarity and recursion of geometric patterns in nature.[14] Eventually, scientists extended the use of the term to describe the self-repeating similarity in growth processes that unfold in time.[15] Yet, no matter whether sound, share price or voltage is involved, the assembly of fractals and the fractal itself ends up being represented by a geometric figure. In a connectio-dynamic worldview, however, shape or geometry is a perceptional topic that does not play a role in the passage of nature. Hence, self-repeating similarity must be found in the process that gives rise to shape, the process of behavior-pattern emergence itself. The ground Gestalt of time with its four quadrants or stages of behavior-pattern emergence is a *fractal of time*. It describes the emergence of reality at all scales. It is recursive and self-similar in that it has the capacity to inspire and drive adjacent realities in each of its four stages. For one, when human organizations traverse a stage of behavior-pattern emergence, they specialize to deal best with the stage-specific challenges. These challenges in themselves are inequalities. The stage-optimized world that these inequalities trigger is a behavioral reality that travels itself through the four stages of behavior-pattern emergence. All in all, the ground Gestalt of time is like a Russian-doll. In each of its four stages, it re-appears in its entirety. Whitehead's "concrete slab or passage of nature" is a fractal of time, indeed.

Through The Eyes Of Nature

The alignment of the four stages of behavior-pattern emergence with the laws of thermodynamics helps reconcile the behavioral worldview with the material worldview. The latter is a view of our world as an assembly of matter, a view that still dominates our thinking today. Yet, at the heart of the behavioral world, the laws of thermodynamics are out of sync with the process that acts at the level of interacting parts. At that level, laws as much as shapes are alien phenomena. Nature simply has no clue of these abstractions. Nonetheless, driven by inequality, it achieves what three laws of thermodynamics predict. So, to further the process of reconciliation, we need to explore how exactly nature realizes the laws of thermodynamics on its own terms. For this purpose, let us assume two identical containers, each filled with the same amount of gas molecules. When you heat one of the two containers, its molecules will move more energetically than the molecules in the other container. Then, by connecting the two containers, the gas molecules are allowed to move freely between them. According to the Zeroth Law, you'd expect the temperature in one container to go down while the temperature in the other increases until reaching equilibrium. The question is what happens at the level of interacting parts? In the container with the cooler gas, the molecules move less energetically and, thus, occupy space less frequently than the more energetically moving molecules in the other container. Hence, the latter will experience less resistance and move to the container with the cooler gas, drawing peer gas molecules in their slipstream. When the energetically moving molecules bump into the molecules of the cooler gas, they transfer part of their motion energy and make these molecules more energetic too. As a result, the temperature of the container with the cooler gas rises. At the same time, it will become more crowded. When the molecules from the container with the heated gas run into an increasingly dense crowd of more actively moving molecules, the least-energy fitting path and flow will reverse at some point. This reversal will take place several times until the flows fail to find least-energy fitting paths. By that time, as the Second Law predicts, the collective of collisions will have increased the haphazard behavior of gas molecules on average. This increased level of haphazardness represents the increase of entropy, the cost of transformation. Assuming that the containers are well insulated and no energy is lost to the external environment, the mix of energy may have changed but not the total energy. This is what the First Law predicts. Lastly, as the Dutch theoretical physicist, Erik Verlinde, suggested, the push exerted by the collective of energetically moving molecules is an entropic force because, in the end state, entropy has increased. In sum, nature is able to act according to the laws of thermodynamics through its three fundamental behavioral roots: *Motion*, *Least-Energy Fitting*, and *Memory* (or repetition). This is all that nature needs to manifest itself.

Dancing Borders

When it comes to the explanation of the behavioral worldview, one fly in the ointment remains. So far, the explanation of how nature manifests itself still hinges on a "dance of interacting parts", parts such as molecules, people, or birds. Of course, no matter whether they are interacting or not, parts belong to the material worldview. So, the last step in the reconciliation of the behavioral and material worldviews involves a reassessment of parts. Does nature really involve parts? If not parts, then what *does* it involve? By and large, borders as the outlines of shape determine the parts that we observe. Despite the flow of liquid molecules that produces a convection cell, its border determines its being. Despite the behavior patterns of people that sustain an organization, its cultural borders determine its existence. And, as the Chilean biologists, Maturana and Varela, stressed, despite the molecules that interact inside a biological cell, the cell wall determines its closure and self-organizing being.[16] Like borders, cell walls allow only certain molecules to pass.[17] As outline, a border can be open or closed. A flock of birds has a border that is open at the back end. A cell wall, on the other hand, is closed because it fully encloses the cell content. Then, what does a border represent? The answer to this question depends on which side of the border you stand. If you stand on the inside of a border then "the outside" is the environment that "the inside" has to deal with. However, if you'd fly over Europe or inspect a cluster of cells through a microscope borders become visible as stretches where different environments meet. At both sides of these stretches, you may find behavioral phenomena at different levels of complexity that traverse a different stage of emergence. Thus, reversing the earlier discussed sequence of dependence, a border essentially is a stretch where inequality manifests.

Inequality ➜ behavior patterns ➜ shape (thus, borders)

Obviously, if an inequality is a behavioral difference of sorts then a border is a virtual stretch where behavior patterns come together and interact. However, again, we appear to have been misled by our observations. When we typically think of borders, we think of visible delineations, such as the delineation between a snowflake and the atmosphere. We know that frozen water molecules vibrate in lattices below the visible shape of a snowflake and that molecules in the atmosphere move, rotate and vibrate. But, because we can't see the interaction, we have come to focus on borders more than on the behavioral contexts that meet. In the light of the latter, the research of the American scientist, George Haller, is revealing in that it clarified the role of borders between *flows*. Haller labeled these borders *Lagrangian coherent structures* after the Italian-born French mathematician, Joseph-Louis Lagrange, who inspired the related mathematics. Haller and peers showed that no matter whether it concerns the behavior patterns of railway passengers, ocean currents or atmospheric turbulences, the invisible dancing borders between these patterns play an important role in the process of nature.[18] The invisible dancing borders between groups of goal-driven passengers shape the flow in a railway station as much as the invisible dancing borders between runway airflows shape the turbulence that disrupts the take off and landing of aircraft.[19]

Haller noted that even though motion lies at the heart of these dancing borders they are surprisingly robust.[20] As discussed in the previous section *Through the Eyes of Nature*, motion also lies at the heart of nature. Consequently, at levels where elementary particles spontaneously emerge, one can expect the interaction of fields to produce borders (or strings), of which the dance pattern shapes the appearance of a particle.[21] Hence, dancing borders rather than parts are the fundamental entities of which the universe is composed. "Intangible and immaterial", dancing borders firmly underpin the idea of reality as immaterial and wholly behavioral phenomenon.

Image Of Revolution

By now, it might be clear that scientists from across many fields have unveiled fragments of the new behavioral worldview. As a result, the material worldview, which has ruled human thinking since Cro-Magnon roamed the continents some 40,000 years ago, may well see its end. [22] A revolution may be in the offing in that the behavioral worldview opens the gateway to behavior-pattern-inspired insights that resolve currently known scientific and technical issues, some of which I have touched upon. When it comes to societal issues, however, the impact of the rise of the behavioral worldview and the decline of the material worldview could really be shattering because the transition is bound to alter the premises of perception and belief. In Part 4, I'll explore the prime shifts that will carry the process of transformation (that may be) ahead of us. One might ask why nobody before pieced together the behavioral worldview? In fact, the man who just about did is often referred to as "one of the most-quoted but least-read philosophers in the Western canon".[23] The British mathematician and philosopher, Alfred North Whitehead, a contemporary of Einstein, unveiled (as I'll explain further on) about 78% of the behavioral worldview. In lectures, compiled in both *The Concept of Nature* and *Process and Reality*, Whitehead presented his novel ideas.[24] Inspired by the French philosopher, Henri Bergson, (his idea of duration) and the Austrian physicist, Ernst Mach, (his idea of including the effect of all the dynamically significant masses in the universe), Whitehead postulated a non-material world that revolves entirely around "events".[25] With an undertone that might hint some irritation, Whitehead referred to the "dogma of materialism", when he questioned the thinking of the scientists of his day, including Einstein, who treated the abstractions of objects, such as points of space and time, as more genuine representations of reality than the relations that produced them. To Whitehead, this was "misplaced concreteness".

> *[I want to shed] light on the changes in the background of our scientific thought, which are necessitated by any acceptance, however qualified, of Einstein's main positions.*

Interestingly, Whitehead's ideas are known for their "staggering complexity".[26] Even his countryman, Julian Barbour, who pictured reality as an odd progression of shapes in a static, predefined spatial world, made similar comments.[27] Considering Whitehead's non-spatial view of reality, this is not surprising. As Kuhn noted, paradigm shifts are not easily accepted, let alone adopted. The following analysis illustrates how the behavioral worldview relates to Whitehead's world. This analysis suggests that his ideas might have been unfamiliar rather than complex.

Non-material world of behavior patterns

In a world made up of behavior patterns, we perceive objects because the behavior patterns that shape these objects are reproducing. Whitehead's version: "The aggregate of events is the solid." Of course, events are not exactly behavior patterns.

So, this is where the views of Whitehead are essentially different from mine.

The false idea that we have to get rid of is that nature is a mere aggregate of independent entities [objects, in other words].

Worldview Features	
Behavioral Worldview	**Whitehead**
Non-material world of behavior patterns	Partly
Nature is process, being is becoming	Yes
Figure Gestalt of time, manmade, repetition	Partly
Simultaneity is at the heart of existence	Yes
Ground Gestalt of time, passage of nature	Partly
Motion, least-energy fitting, memory	Partly
Time, space as emergent phenomena	Yes
Passage of nature as fractal of time	Yes
Dynamically connected realities	Yes
Estimated % identified[28]	78%

Table 2 How the behavioral worldview relates to Whitehead's ideas.

Nature is process, being is becoming
The behavioral worldview depicts nature ultimately as a *process that unfolds* from one instant to the next. This fully resonates with Whitehead's views. In the light of this, Whitehead referred to the ideas of the Russian philosopher, Nikolay Lossky.[29]

A mature philosopher raises nature to independence and makes it construct itself and he never feels, therefore, the necessity of opposing nature as constructed.

Figure Gestalt of time, manmade, repetition
In the behavioral worldview, time is manmade and a measure of repetition rather than change. Whitehead was on the same line

of thinking when it comes to the origin of time. Yet, he didn't define it. "Time is a metaphysical enigma" (or puzzling).

Serial time is the result of an intellectual process of abstraction. It is evidently not the passage of nature itself.

Simultaneity is at the heart of existence
In the behavioral worldview, as behavior patterns reproduce, order and, thus, existence hinges on simultaneity. In the mind of Whitehead, simultaneity was crucial too. He referred to it as "temporal congruence" or harmony from a timing viewpoint.

Simultaneity is a definite natural relation. A certain whole of nature is limited only by the property of being simultaneously. Simultaneity is the property of a group of natural elements, which in some sense are components of "duration" [passage of nature].

Ground Gestalt of time, passage of nature
Whitehead coined the term "the passage of nature", and referred to it as the other, more basic aspect of time, building on the idea of "duration", which Bergson had identified. Redefined as the *ground Gestalt of time* with four stages or *quarters* of behavior-pattern emergence, it is imperative to the behavioral worldview. The passage of nature essentially is the life cycle of (congruent) simultaneity. For reasons discussed next, Whitehead did not specify four stages. However, in *Process and Reality*, he did paint them albeit with broad strokes.

A cosmic society arises from disorder where disorder is defined by reference to the ideal for that society; the favorable background of a larger environment either itself decays, or ceases to favor the persistence of the society after some stage of growth: the society then ceases to reproduce its members, and finally, after a stage of decay, passes out of existence.

137

Motion, least-energy fitting, memory

Whitehead believed that "motion is an axiom" or self-evident. Yet, he missed out on the other fundamental behavioral roots: the *Least-Energy Fitting principle* and *memory* (repetition). Of course, he had not quite made the jump from *objects* (as players in events) to *behavior patterns* (triggered by inequalities). This produced two limits to his views. First, as Whitehead admitted, he did not quite resolve the question of causality, which, in the behavioral worldview, depends squarely on the *sequence of dependence* and on the energy conversion involved.

Inequality ➔ behavior patterns ➔ shape

In *Process and Reality*, Whitehead noted his sentiments on this.

> *It is evident that the ingression of objects into events includes the theory of causation. I prefer to neglect this aspect of ingression because causation raises the memory of discussions based upon theories of nature [religion, probably], which are alien to my own.*

Then again, Whitehead acknowledged the significance of *congruence*, defining it as a blend of *spatial congruence* (spatial harmony) and *temporal congruence* (simultaneity) or, in short, *congruent simultaneity*. He went as far as predicting the coming of "a theory of temporal congruence". However, not having made the switch from events to behavior patterns, the stages of behavior-pattern emergence remained beyond his reach.

> *Uniformity in change is directly perceived, and it follows that mankind perceives in nature factors from which a theory of temporal congruence can be formed.*

(Ordinary) time, space as emergent phenomena

In the behavioral worldview, both ordinary time (*figure Gestalt of time*) and shape emerge by the *sequence of dependence*. As a measure of repetition, ordinary time emerges once behavior patterns start reproducing. Subsequently, shape emerges from the repeated dance of behavior patterns involved. In the end, as Whitehead emphasized in *The Concept of Nature*, space follows shape as a measuring convention (height, length, width).

> *Spatial order is derivative of temporal order. Space and time are merely ways of expressing certain truths about the relations between events. Space is not a relation between substances but between attributes.*

Passage of nature as fractal of time

In the behavioral worldview, no matter whether it concerns ant behavior, phase transitions (between gas, liquid, solid states), or the development of firms, nations and economies, the passage of nature or *ground Gestalt of time* and its four stages consistently appear throughout. Moreover, each stage of emergence involves the entire passage of nature. Whitehead, on his part, confirms the fractal quality of nature in terms of events being nested, each event involving the "concrete slab or passage of nature".

> *The continuity of nature arises from extensions. Every event extends over other events and every event is extended over by other events [nested, in other words].*

Dynamically connected realities

In the behavioral worldview, behavior patterns emerge locally in response to inequalities that themselves have been induced by behavior patterns on different paths of nature. Hence, when behavior patterns weave the realities that we observe this way, they are essentially connected. In line, Whitehead pictured our world as a world of connected events rather than a world of "independent entities", such as objects.

> *For us the red glow of the sunset should be as much part of nature as are the molecules and electric waves by which men of science would explain the phenomenon. It is for natural philosophy to analyze how these various elements of nature are connected.*

Discussion

If Whitehead had identified a world of behavior patterns rather than events, he would have come full circle. In view of the state of physics at the time, Whitehead's focus on events probably was a compromise between the connected, constantly emerging world that he imagined and the views of the physicists of his day who had acknowledged spacetime events. Events, thus, seemed like a "legitimate" replacement of objects. So, not just scientists but also (natural) philosophers, such as Whitehead, Bergson, Mach, Lossky, and Prigogine, were on the verge of unveiling the behavioral worldview.

One curious aspect of this worldview is that humanity itself becomes nature's often-unwilling (as it sometimes seems) behavioral device that helps it convert energy more and more efficiently because, in the end, that is what nature, *as process*, is about. The behavior patterns that make up society are subject to the least-energy fitting principle too and, thus, we can expect to see the emergence of new insights and patterns that will allow us to fulfill the increasing expectations of nature. Another curious aspect, as we may come to realize, is that *man* is not necessarily "the chosen one" but a transitional behavior-pattern-inspired phenomenon on the path of nature that nature itself follows. A less peculiar, but nonetheless intriguing aspect is that the path of nature and its four stages of emergence become visible in every phenomenon of reality once we know they exist. As a result, our world will become more predictable as one stage of emergence is sure to be followed by the next stage. In other words, we are riding in a rollercoaster with four predictable bends that cannot be avoided. No matter which strategies we follow, economies, nations and companies will traverse four stages of rising and declining growth. We might be able to slow down these societal phenomena when traversing

140

stages of rising and stable growth and to accelerate when traversing the stages of declining and uncertain growth. This seems possible because, knowing the stage-specific situation, we might invent the necessary strategies and actions. One curious aspect remains. If economies, nations, companies, *and* physical phenomena are behavioral phenomena then so are the "societies of mind". Hence, in line with nature's *sequence of dependence*, "shape" (thoughts, choices) will arise in the mind from the behavior patterns of neuron clusters only.[30]

Inequality ➔ behavior patterns ➔ shape (thought, choice)

Yet, before these behavior patterns develop, inequalities prompted by "the senses" or by behavior patterns of neuron clusters on other paths of nature will have inspired them. So, without losing our experience of free will, we will come to realize that our decisions reflect the turns that neuron clusters have made *on our behalf* stimulated by local inequalities that lie beyond our awareness. The French philosopher, Henri Bergson, who unveiled the passage of nature first (labeling it "duration"; not the kind of duration that we usually refer to), understood this all too clearly. [31]

In cases where action is freely performed, we cannot reason about it without knowing the conditions that caused it. If we pronounce our actions as free, this is so because we do not understand the relation of this action to the state from which it is issued.

As the American neuroscientist, Sam Harris, pointed out much later, research showed that "conscious decisions could be predicted up to 10 seconds before they enter awareness." So, although we appear to be aware of our choices, Harris reminded his readers, "We are not the authors of our actions."[32] In a nutshell, inequalities rather than "we" determine what we think and choose. However thought provoking this might be, this is just the top of the iceberg. In the next part, I'll explore other shifts in science and society that the behavioral worldview will bring about once it finds traction. I will try to find an answer to the following question: *From a scientific, professional, and moral point of view, how will the behavioral worldview change our lives?*

Notes

[1] Julian Barbour, *The End of Time: The Next Revolution in Physics*, Oxford University Press (1999).

[2] As Plato already identified in his book, *The Republic*, the origin of a nation "*lies in the natural inequality of humanity that is embodied in the division of labor*".

[3] Lars Onsager, *Reciprocal Relations in Irreversible Processes* I., Physical Review No. 37, pp. 405-426 (1931).

[4] The former is the so-called *Peltier effect* and the latter the so-called *Seebeck effect*.

[5] Ilya Prigogine, *Time, Structure, and Fluctuations*, Nobel Lecture (1977).

[6] Rod Swenson, T*he Fourth Law of Thermodynamics or the Law of Maximum Entropy Production*, Chemistry, Vol. 18, Issue 5 (2009). In the quoted text, I changed "potential" into "inequality" not to confuse the reader.

[7] See Part 2, *The Alley of Time.*

[8] Any translational energy (rectilinear movement) is shared with the other frozen water molecules too.

[9] The British philosopher, Alfred North Whitehead, summarized "congruent simultaneity" in one term: "congruence". *"Congruence depends on motion and is thereby generated by the connection between spatial congruence [congruence] and temporal congruence [simultaneity]."* Alfred North Whitehead, *The Concept of Nature* (Lectures delivered in 1919 at Trinity College, Cambridge), Cambridge University Press (1920), p. 137.

[10] The rotation and translational energy of settling molecules may temporarily unfreeze frozen molecules at the surface of an ice crystal: "*Surface melting has a profound effect on the surface structure of ice, and attachment kinetics undoubtedly depends on surface structure. Thus we expect that surface melting must play an important role in the growth of snow crystals. Two additional facts support this: (1) surface melting is known to depend strongly on temperature over just the range where snow crystal growth shows a great deal of variation with temperature, and (2) different facets of the same solid sometimes exhibit the*

effects of surface melting differently. Thus it may well be that much of the temperature variation seen in the morphology diagram is a manifestation of surface melting in ice." Kenneth Libbrecht, *The Physics of Snow Crystals*, Institute of Physics Publishing, Reports on Progress in Physics, 68 (2005) pp. 855-895. http://www.its.caltech.edu/~atomic/publist/rpp5_4_R03.pdf

[11] Tsuda refers to these distinctive sequences as "temporal codes" because they represent the (entirely) time-inspired firing behavior of neurons. Kunihiko Kaneko, Ichiro Tsuda, *Complex Systems: Chaos and Beyond*, Springer-Verlag, Berlin (2000).

[12] Particularly in the first three chapters of the first book: Adam Smith, *An Inquiry into the Nature and Causes of the Wealth of Nations*, London (1776).

[13] *How rocks evolve*, The Economist (13 November 2008). Probably based on the following article: Robert Hazen et al, *Mineral evolution*, American Mineralogist 93, pp. 1693-1720 (2008).

[14] Benoît Mandelbrot, *The fractal geometry of nature*, W.H. Freeman, New York (1983).

[15] Tamás Vicsek, *Fractal Growth Phenomena*, World Scientific Publishing, Singapore, London (1992).

[16] Humberto Maturana, Francisco Varela, *Autopoiesis: The Organization of the Living*, D. Reidel Publishing Company, Dordrecht (1972).

[17] The cell wall of a biological cell is a membrane that allows the passage of certain molecules only. Typically, molecules that serve as resources or food are allowed to enter and molecules that need to be wasted are allowed to leave.

[18] *The Skeleton of Water*, The Economist (14 November, 2009). George Haller, *Chaos Near Resonance*, Springer, New York (1999).

[19] Wenbo Tang, Pak Wai Chan, George Haller, *Lagrangian Coherent Structure Analysis of Terminal Winds Detected by Lidar. Part II: Structure Evolution and Comparison with Flight Data,* Journal of Applied Meteorology & Climatology, Vol. 50, Issue 10, pp. 2167-2183 (October 2011).

[20] George Haller, *Lagrangian coherent structures from approximate velocity data*, Physics of Fluids, Vol. 14, Number 6 (June 2002).

[21] I refer here to the common assumption in physics that the oscillation behavior of a string determines its role or image as elementary particle.

[22] Currently referred to as European Early Modern Humans (EEMH). I continued referring to "Cro-Magnon" because I did not want to leave out today's non-Europeans who all stem from Cro-Magnon too.

[23] Gary Herstein, *Alfred North Whitehead (1861-1947)*, IEP (8 May 2007). http://www.iep.utm.edu/whitehed/

[24] Alfred North Whitehead, *The Concept of Nature* (Lectures delivered in 1919 at Trinity College, Cambridge), Cambridge University Press (1920) and Alfred North Whitehead, *Process and Nature*, The Free Press, New York (1978).

[25] It is perhaps worth mentioning here that the British mathematical physicist, Roger Penrose, eventually introduced the "*twistor* theory", which depicts a world made up of *events*. Roger Penrose, Fedja Hadrovich, *Twistor Theory*, http://users.ox.ac.uk/~tweb/00006/index.shtml Smolin later reported that Penrose's twistor theory, however promising, was eventually viewed as not *causal* enough. Lee Smolin, *The Trouble with Physics*, Houghton Mifflin, Boston-New York (2006).

[26] In the introduction: Gary Herstein, *Alfred North Whitehead (1861-1947)*, IEP (8 May 2007). http://www.iep.utm.edu/whitehed/

[27] Near the end of Barbour's book: Julian Barbour, *The End of Time: The Next Revolution in Physics*, Oxford University Press (1999).

[28] If each "Yes" = 1 and each "Partly" = 0.5 then this amounts to 7/9 or 78%.

[29] Nikolay Lossky, *The Intuitive Basis of Knowledge* (original in 1903, translation in 1919).

[30] I discussed this in Part 2 in *The Alley of Complex Systems*.

[31] Henri Bergson, *Time and Free Will*, Dover, London (2001).

[32] Sam Harris, *The Moral Landscape*, Free Press, New York, 2010.

Prime Shifts

The behavioral worldview, its sequence of dependence (which essentially is about the cause of the realities that we observe), and the predictive capacity of the four stages of emergence can be seen to influence the thinking and acting in a broad range of fields, from the natural sciences, business sciences and morality to the material sciences and the evolution of share price. The question that I'll address here is how the behavioral worldview might affect our everyday lives. There is no other agenda. With the target audience of this book in mind, I have grouped the 20 trends that I identified under the following two headers:

      i.   *Science-related* trends and
    ii.   *Society-related* trends.

Of course, because the nature of reality is involved, the effect of some trends will be felt in both groups.

Science-Related Trends
Science-related trends influence the course of scientific research and/or alter the philosophical underpinning of science.

1 - From objects to behavior patterns
Considering Einstein's Theory of Relativity, which explains that energy and matter are transmutable, the crux of reality today is in the materialization of energy. Thus, the smallest entity of which the universe is made up is an object or particle. However, when the behavioral worldview finds traction, objects and particles become viewed as temporary manifestations of reality that are visible to our senses only because the behavior patterns that shape them are reproducing long enough to be observed. We will see a universe made up of dancing borders in-between differing local environments. The transition from material to behavioral worldview will also help explain why particles can be observed in more than one location at the same time.[1]

Inequality ➔ **behavior patterns** ➔ shape (visible particle)

As reproducing behavior patterns, particles don't just occupy one location but a virtual dance floor. Depending on the shutter speed, we may record one and the same particle at both ends of the dance floor "at the same time".

2 - From mutual causality to reciprocal conditioning
A world of things or particles is rigid in the sense that one effect typically follows another over time. For example, an event, involving *particle A*, might influence another, subsequent event, involving *particle B*. After all, particle A and B are distinct or, as Whitehead said, "isolated" things in some spatial context. In the material worldview, therefore, causality is a serial phenomenon. Yet, very recent insights into quantum mechanics suggest that this doesn't always apply at levels where elementary particles wander.[1] Particle A might influence the rise of particle B as

much as particle B influences the rise of particle A. Such cases of "mutual causality" cannot easily be explained in the material worldview because they break the principle of seriality. At best, as the report stated: "A causes B causes A". But, that is tricky because it reduces mutual causality to a circular "chicken and egg" question: Which is the primary cause? Indeed, as hinted before, in the material worldview, causality is an issue that confused even Whitehead. The behavioral worldview sheds new light on this question. As the *sequence of dependence* (below) dictates, particles are emergent behavioral phenomena rather than isolated objects. Their appearance or shape is but the observable fruit of reproducing behavior patterns. Hence, in the behavioral worldview, the reality of an elementary particle is much broader than what the material worldview assumes.

Inequality ➔ behavior patterns ➔ shape (visible particle)

Moreover, the sequence of dependence shows how one spark may trigger the emergence of two or more particles at the same time (Figure 16). As discussed in Part 3, behavior patterns may function themselves as inequalities that give rise to yet other behavior patterns. However, the rise of one can be a condition for the rise of another *at the same time* when the rise of another is a condition for the rise of one. This process basically rests on *reciprocal conditioning*.[2] Whereas *causality* is a more active and serial agent, *reciprocal conditioning* is passive and simultaneous. Gautama Buddha was one of the first thinkers to acknowledge this when he said: "Only together [realities] arise."[3] Unsurprisingly, reciprocal conditioning also explains the passive and simultaneous rise of societal phenomena, which hinge on behavior patterns too. The rise of *the individual* is a condition for the rise of *society*. Yet, at the same time, the rise of *society* is a condition for the rise of the *individual*. The rise of *autonomy* is a condition for the rise of *inter-dependence*. Yet, at the same time, the rise of *inter-dependence* is a condition for the rise of *autonomy*. More controversially, the rise of *the idea of God* is a condition for the rise of *religion*. Yet, at the same time, the rise of *religion* is a condition for the rise of the *idea of God*.

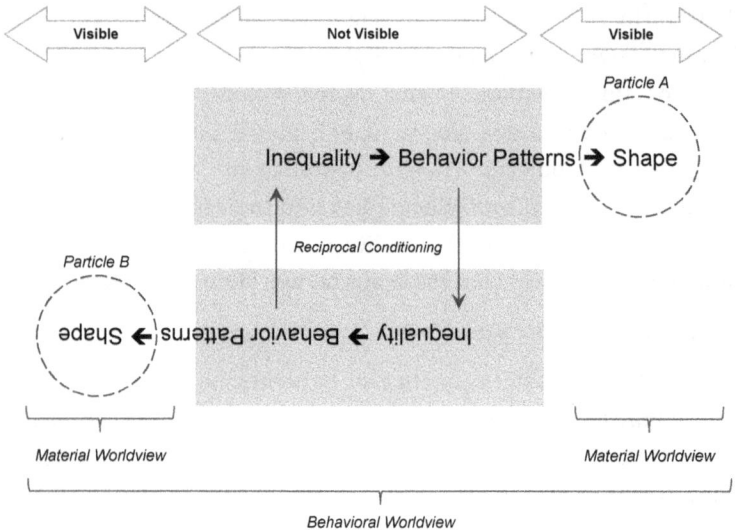

Figure 16 Through reciprocal conditioning, only together do they arise.

3 - From diversity to uniformity

Today, the diversity of observable realities produces "awe and wonder" in creationists and evolutionists alike. However, in the behavioral worldview, all that we observe (and not observe) can be reduced to emergent behavior patterns. Therefore, when the behavioral worldview finds traction, the "awe and wonder" of laymen and scientists alike will not be for the diversity of nature but for its uniformity.

4 - From natural selection to least-energy fitting

When it comes to *natural selection* or *survival of the fittest*, the German evolutionary scientist, Georgy Koentges, observed: "the central problem is the impossibility of defining fitness, whether in organisms, organs, cells, genes or even gene regulatory DNA regions."[4] Wallace and Darwin may have identified the principle of natural selection but this principle does not at all make clear how the process of natural selection unfolds. It only predicts the end result: Best-fitting species out-reproduce those that don't! Hence, as Koentges' observation suggests, causality remains a problem in need of solution. *What exactly does it mean to be the better-fitting species?* As I wrote in *A New Leadership Ethos*, the British evolutionary biologist, Richard Dawkins, showed that an extensive analysis is needed to identify the traits that make a species best fitting given the environmental conditions.[5]

In [his book] "Climbing Mount Improbable", Dawkins explores in detail the path of natural selection toward increasingly more complex life forms. Chance emerges as an important attribute of the self-propelled process of evolution because it explains the remarkable diversity of species. However, apart from chance and the overarching principle of the survival of the fittest, each case of species-development stands on its own and requires extensive analysis, if not speculation as to why certain traits may have been favored over others. [Furthermore,] in "The Extended Phenotype", Dawkins explores a hypothetical case to show why nature favored the development of functional units over one organism that does it all. With a sense of detail and mastery that is sure to have thrilled peer biologists, Dawkins essentially gives an account of the division of labor at the level of molecules and cells, illustrating how such a division serves the preference of nature to seek the most efficient solution in constantly changing circumstances.

As process, nature cannot mimic Dawkins' mastery when it comes to identifying best-fitting species. However, according to the behavioral worldview, it works at a level where it blindly achieves the same result. At the level where behavior patterns rule, least-energy fitting behavior patterns passively survive or out-reproduce behavior patterns that lose more energy to the environment (for example, by interfering with others too much). Thus, as a process that involves *motion, least-energy fitting* and *memory* (or repetition), nature has causality at its center. What's more, nature's passive and *local* reliance on least-energy fitting patterns takes "chance" out of the equation. As the *sequence of dependence* shows, behavior patterns arise simply when inequalities occur. Chance is not needed. As I noted in *To Be Or To Become*, quite recent research confirmed that even an exceptionally stable hierarchy of behavior patterns such as the DNA molecule "adapts" to its environment by "epi-genetics".[6] In a nutshell, considering that all cell types contain the same DNA, epigenetics explain how *local* inequalities either silence or activate sequences of DNA to help cells differentiate and grow into the different body cells. These local inequalities may even be influenced by behavioral changes, such as a weather-inspired diet change. In sum, as thermodynamic and causal phenomenon, the *least-energy fitting principle* will push *natural selection* from its much-debated pedestal. Curiously, of the two, *least-energy fitting* is the closest thing to Adam Smith's "invisible hand".

5 - From divergent to convergent evolution

A frequently discussed controversy among paleontologists, paleobiologists and other evolutionary scientists involves the chance-driven nature of "natural selection". Chance supposedly ensures diversity because it (supposedly) is random. What is more, because chance relies on no other criterion but chance, the outcome of an evolutionary process cannot be predicted. In practice, this means that the evolution of life on Earth might take a different path should it be rebooted. Dawkins and the

American paleontologist, Stephen Jay Gould, strongly defended this line of thinking.[7] However, with the introduction of the behavioral worldview, in which *least-energy fitting* takes over from chance as the principle of selection, the idea that evolution is divergent can no longer be maintained. Unlike chance, least-energy fitting essentially represents a local energy optimization criterion that nature follows throughout. Because least-energy fitting behavior patterns are consistently favored, the outcomes produced by evolution are of necessity convergent rather than divergent. In his book *Life's Solution*, the British evolutionary paleobiologist, Simon Conway Morris, identified numerous examples where nature arrives at common solutions whenever a new branch emerged on the tree of species and regardless of the direction that history took.[8] For one, Conway Morris showed how nature's solutions in mammals *converged* to common traits, such as two eyes, four legs, one mouth closely positioned near the brain, high oxygen-carrying capacity of the blood, a single aorta and seven-neck vertebrae. Common societal or behavioral solutions echo throughout the course of history too. This might explain why leaders today are still inspired by the tactics of an ancient Chinese warrior and why history seems to repeat itself. In all, when the behavior worldview finds traction, the idea that evolution is convergent rather than divergent will prevail.

6 - From creation to self-creation

The ethos today is that we are creators not unlike the God(s) that we worship. Leaders believe they can "create" companies as much as engineers and alchemists believe they can "create" materials. In the behavioral worldview, this cannot seriously be maintained. Only inequalities inspire the emergence of realities through the behavior patterns that they evoke. Hence, rather than creating companies, leaders can set the conditions for their spontaneous emergence, for example, by introducing reward-related inequalities (assuming the market conditions are right). Following essentially the same recipe, scientists grew a material with trillions of vacuum holes, which increased the material's insulating capacity.[9] These holes spontaneously emerged when the engineers tinkered with the environmental conditions as they grew the material. By using this "self-assembling" material as substrate for microchips, the wiring can be etched closer together to gain speed and reduce energy consumption without inviting unwanted cross-wire electrical leakages. So, to achieve and sustain self-creation, the behavioral worldview will direct our focus to the management of environmental conditions.

7 - From equilibrium to non-equilibrium thermodynamics

As discussed in some detail in Part 3, realities emerge from the surface of the equilibrium ocean to reach the domain of non-equilibrium thermodynamics where nature relies on the *least-energy fitting principle* and *reproducing behavior patterns*. Hence, once the behavioral worldview becomes adopted, the eyes of more and more scientists will settle on the question of non-equilibrium thermodynamics in search of behavior-pattern-inspired solutions to known material-worldview limits. An early example involves the partially successful efforts of a very young American researcher who hopes to reach beyond the accepted limits of equilibrium thermodynamics when it comes to splitting water into hydrogen and oxygen. The thinking of this researcher shows an innate sense of the behavioral worldview.[10]

> *Everything you see is shaking. Solidity is an illusion. Quantum physics explores the vast inner universe of molecules, the open spaces where atomic particles vibrate and struggle against one another in waves. Kinetic energy is in everything! The clue is in the resonant frequency of molecules.*

8 - From Figure Gestalt of time to Ground Gestalt of time

Indirectly acknowledged by Einstein's prediction that nothing can travel faster than light, time has had a strange attraction on scientists and laymen alike. Some want to travel to the future, some to the past. Some want to reverse it, while others claim that that can't be done.[11] Some explain that time is not needed; it's a figment of the human mind. Yet, by and large, the material worldview depends on it. In the behavioral worldview, however, ordinary time is a measure of repetition. Because repetition is what makes behavior patterns visible, it represents the figure Gestalt of time. The true measure of nature unfolding is in the ground Gestalt of time and its four universal quarters or stages of behavior-pattern emergence. So, this is the time to reckon with when the behavioral worldview spreads.

9 - From Holon to Tilon

As noted in Part 1, in *The Ghost in the Machine*, the Hungarian-born British journalist, Arthur Koestler, coined the term "holon" to refer to something that is both *a whole* and *a part*.[12] By identifying a single term for every observable phenomenon of reality, Koestler effectively summarized our world in one word. In the behavioral worldview, however, reality is a much broader phenomenon, which includes all four stages of behavior-pattern emergence, not just the stage(s) in which reality is observable. As fractal of time, these four stages appear at every level of our world not unlike a "holon". Together, as quarters on a clock that displays the ground Gestalt of time, they therefore represent a "time-holon" or "tilon". Accordingly, the "hierarchy of tilons" or "tilarchy" depicts the simultaneous unfolding of nature at all levels. When the behavioral worldview is adopted, this time-based hierarchy will dominate the explanation of our world.

10 - From dualism to monism

As the 17th century French philosopher, René Descartes, wrote, reality involves both matter and thought. Through this dualism, Descartes effectively served the scientific thinking of his time in that he allowed contemporaries, such as Huygens and Newton, to develop explanations of the material world in isolation from religious thinking. However, in the behavioral worldview, mind and matter become unified in that they are reduced to emergent behavior patterns. This reduction is likely to foster a shift from "dualism" to "monism", a shift that seems timely because, as Kuhn noted, "in science [fragmented as it is today], reduction is desirable because it explains why and how something exists."

Society-Related Trends

Society-related trends influence the way we think, act, and lead in day-to-day productive human life.

11 - From equality to inequality

Historically, the disproportionate distribution of wealth has directed the focus of societal leaders across the world to the minimization of inequality. "Equality" now lies at the heart of human society. Yet, however virtuous equality might be, it is in the way of an adequate appraisal of the process of nature. The path that leads to equality unavoidably starts with inequality. According to the behavioral worldview, no reality arises without inequality arising first. Inequality, thus, is more fundamental to (our) existence than equality. What does this mean for equality as virtue? When the behavioral worldview becomes adopted, equality will be seen as the virtue of action rather than inaction. In other words, the experience of equality should be found in the minimization of inequality (and, thus, in "work"). As a result, the idea of inequality, as the root of existence, will gain in importance at the cost of the idea of equality.

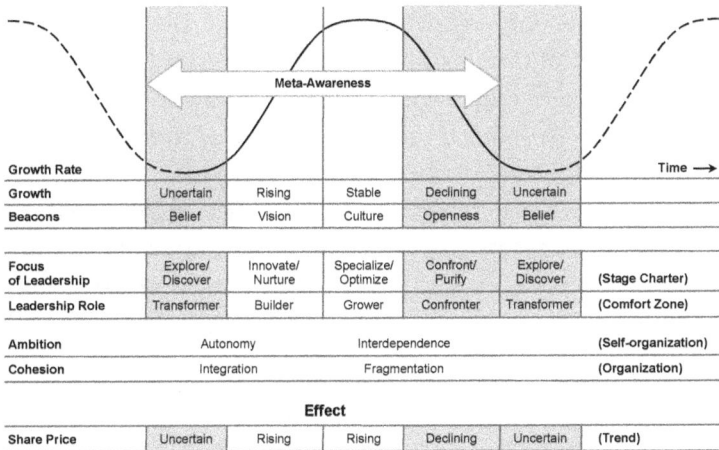

Growth Rate					Time →→
Growth	Uncertain	Rising	Stable	Declining	Uncertain	
Beacons	Belief	Vision	Culture	Openness	Belief	

Focus of Leadership	Explore/ Discover	Innovate/ Nurture	Specialize/ Optimize	Confront/ Purify	Explore/ Discover	(Stage Charter)
Leadership Role	Transformer	Builder	Grower	Confronter	Transformer	(Comfort Zone)

Ambition	Autonomy		Interdependence			(Self-organization)
Cohesion	Integration		Fragmentation			(Organization)

Effect						
Share Price	Uncertain	Rising	Rising	Declining	Uncertain	(Trend)

Figure 17 A differentiated, temporal understanding of organization.

12 - From awareness to meta-awareness

Apparently still inspired by Descartes' mechanistic approach to reality (the material worldview), leaders generally treat nations, organizations, and even economies as mechanical contraptions that can be repaired by changing some nuts and bolts (people, tools) and/or by changing the way they function (processes, procedures). As discussed in Part 2, distinguished management gurus and top-of-the-bill consultants have a habit of reinforcing the idea by suggesting that organizations can be fixed whenever "broken" through strategies that appeared to have worked in the exemplar companies of the day. Apart from hope, however, they seed disaster. First, the exemplar companies of the day by and large turn out to be the losers of tomorrow. Second, when prepped-up leaders invest time and money to prolong their organization's winning streak, they often discover that they are "flogging a dead horse". Third, time and again, societal leaders are overwhelmed by the sudden collapse of their organization, nation, or economy because they fail to see the unavoidable consequences of their cyclical nature. Offering a differentiated, temporal perspective (Figure 17), the behavioral worldview helps leaders develop "meta-awareness", that is, an awareness of the temporal state of their organization, nation, or economy. Hinging on emergent behavior patterns, organizations, nations, and economies necessarily traverse the four stages of behavior-pattern emergence, each with specific challenges and solutions. In future, leaders will have no reason to ignore this.

13 - From timeless leadership to temporal leadership

The current ethos of leadership doesn't limit the reign of leaders as a rule. Successful leaders often assume they can manage an organization or nation no matter what its state of development. Yet, top-level executives who excelled in General Electric under the ardent grower, Jack Welch, failed to achieve success in companies that went through a different stage of development.[13] When push came to shove in these companies, they opted for measures that fit their comfort zone. Because Welch had made sure to hire innate "growers", these leaders, conditioned by trillions of neural pathways, did what they were familiar with. As the four stages of behavior-pattern emergence show, the challenges and measures in each stage are quite different (Figure 18).[14] Hence, when leaders instill measures that fit their comfort zone, they may either pull an organization into the next stage or, having stayed on too long, prevent it from moving forward. As history has shown, the latter means sliding back.[15] Hence, to ensure that organizations, nations, and economies progress through successive stages of development, temporal leadership is needed. Temporal leadership requires leaders and their management teams to have the comfort zone needed to move the organization forward.[16]

TRANSFORMER
- Introduces a new thinking
- Driven by conviction and belief
- Aims to re-invent organization or society
- May have to make several attempts
- Identifies a new platform for growth

BUILDER
- Turns concept into product
- Driven by vision and passion
- Builds basic organization
- Identifies niches, cozies up to clients
- Proves business through growth

GROWER
- Repeats proven success evermore efficiently
- Driven by cost, volume and quality
- Fosters process, flows and culture
- Fans out responsibilities to speclialists
- Realizes stable growth

CONFRONTER
- Confronts complacency
- Distances from established procedures
- Simplifies and purifies organization
- Reintroduces outside standards of success
- Achieves profitability but not necessarily growth

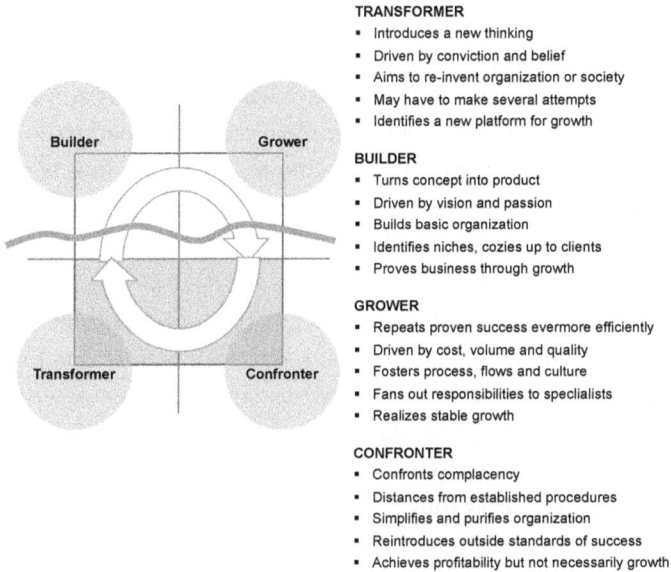

Figure 18 Temporal leadership hinges on leadership comfort zones.

14 - From "Built to Last" to "Grown to Achieve"

The four fundamental stages of behavior-pattern emergence at the heart of the process of nature simply rule out the possibility that anything can be built to last. Changes outside and on the "dance floor" determine the fate of emerging behavior patterns. At best, leaders may slow down or speed up the progression of their organization, nation, or economy when it traverses a certain stage. Of course, in the behavioral worldview, behavior patterns will continue only as long as they haven't minimized the inequalities that produced them in the first place. As a result, organizations, nations, and economies arise spontaneously to minimize certain inequalities then to disappear, fall apart, or re-invent themselves when this has been accomplished. Leadership in the future will be configured on this realization.

15 - From due diligence to future diligence

Measures of organizational success are usually anchored in the past. For example, when accountants are commissioned to do a "due diligence", they determine whether an organization is what the balance sheet says it "was". However, not without reason, the financial services world is required to include in its brochures the statement "*past performance does not guarantee future performance*". Taking into account that organizations are emergent behavioral phenomena and that leaders have comfort zones that generally relate to one particular stage of emergence only,[17] it is not at all certain that an organization will perform in the future as it did in the past. Whereas a leader may have led an organization to success thus far, he or she may lack the comfort zone needed for the next stage. A "future diligence" appraisal ensures that organizations have the strategic ingredients and leaders needed for the next stage of development. It involves the identification of the current and future stage of organizational emergence and an evaluation of the comfort zones of the incumbent leader and his team.

16 - From managing time to managing simultaneity

So far, time has been at the heart of our sense of achievement. The common perception of organizational performance usually includes "being in time". After all, time is money. Yet, to be in time, organizations must achieve simultaneity first. Simultaneity requires people and organizations to work together both at the same time and spatially harmoniously. (According to the *least-energy fitting principle*, this means *congruent simultaneity.*) In the behavioral worldview, therefore, "simultaneity" rather than "time" is what counts when making sure that organizations, people, and contraptions get things done in time.

17 - From timeless practices to temporal practices

So far, best practices, management consulting, and leadership education are designed to ensure success and resolve problems irrespective of the four stages of behavior-pattern emergence.[18] In the material worldview, organizations are typically viewed as machines that can be optimized or repaired to keep them running by means of some standard practice. In the behavioral worldview, best practices, management-consulting services and leadership education are made more potent and to the point by configuring them based on the needs in the different stages of behavior-pattern emergence (Figure 17). The term "temporal configuration" is likely to enter the leadership vocabulary.

18 - From prediction to projection

When the behavioral worldview finds traction, our world will become more predictable. Prediction becomes projection in that the actual behavioral features of corporations, nations, and economies can be projected on the four-stage cycle to identify the current and *future* stage of behavior-pattern emergence. This way, voters and policymakers will be able to evaluate the state of their nation and the fit of candidate leaders. They are likely to be supported by social media that will help chart the comfort zone and fit of leaders by aggregating the opinions of their users.[19] The world of financial services will be affected too. Thus far, stock picking turned out to be closer to betting than investing. Now, the four-stage cycle of behavior-pattern emergence offers a share price trend for each of the four stages based on the growth rate in each stage (Figure 17). Thus, in future, investment strategies will necessarily involve such an analysis. The actual behavioral features of companies will be projected on the four-stage cycle to identify the current stage of behavior-pattern emergence and the share-price trend for this stage will help interpret the ongoing share-price fluctuations. [20]

19 - From free will to sense of destiny

As the French philosopher, Henri Bergson, already suggested, our sense of free will is not proof of having a free will.[21] As the sequence of dependence shows, (local) inequalities determine our decisions and choices.

Inequality ➔ (neuron-cluster) **behavior patterns** ➔ thought, decision, choice

The decision or choice that we think we make has already been made on our behalf at neuron-cluster levels. Hence, each choice represents but a sense of destiny. So, while man is an independent agent with a free will in the material worldview, he is but a device of nature in a wholly behavioral world.

20 - From being to becoming

In the material worldview, an observable object exists or "is". In the behavioral worldview, an observable object represents but a stage of emergence, the observable feat of a constant becoming that is riding on a continuum of inequalities, a mere part of the spectrum of the process of nature.

Discussion

These prime shifts will give rise to secondary shifts across the sciences.[22] On the whole, the focus can be expected to shift to the study of phenomena as complexes of 3-dimensional dance patterns. As a result, a broad range of fields, from nuclear-fusion research to material design, may come to rely on some least-energy fitting "choreography". The scope of these secondary shifts will be such that each may take multiple publications.

We effectively live in an era that was started by the Dutch lens grinder, philosopher, and rationalist, Baruch Spinoza.

Spinoza's era began when his book *Ethica* was published upon his early death in 1677.[23] The behavioral worldview confirms many of the ideas that Spinoza trusted to paper. According to one summary,[24] he offered an alternative to materialism (Trend 1).[25] He identified the unity and regularity of all that happens (Trends 3, 20). He believed that this unity would annihilate the manmade divide between mind and matter (Trend 10). He found the process of nature in an indifferent God (Trends 2, 6, 19). Spinoza even identified the "sequence of dependence".[26]

(External cause) ➔ (self-creation, LEF) ➔ (individual thing, choice)

On "external cause":

Every individual thing, or everything which is finite and has a conditioned existence, cannot exist or be conditioned to act, unless it be conditioned for existence and action by a cause other than itself, which also is finite, and has a conditional existence.

On "self-creation":[27]

A substance cannot be produced by anything external to itself. It must, therefore, be its own cause, that is, its essence necessarily involves existence, or existence belongs to its nature. No substance can be produced or created by anything other than itself.

On LEF or "least-energy fitting":

I think I have shown that from infinite nature, all things have necessarily flowed forth in an infinite number of ways, or always flow from the same necessity. [Nature] is the efficient cause not only of the existence of things but also of their essence. There is no need to show at length that nature has no particular goal in view and that the final causes are mere human figments. Reality and perfection I use as synonymous terms.

On "free will or choice":

Will cannot be called a free cause, but only a necessary cause. [Nature] does not act according to freedom of the will. For, will, like the rest, stands in need of a cause, by which it is conditioned to exist and act in a particular manner.

From the above it may be clear that Spinoza was one of the founders of the behavioral worldview.[28] Lens grinding has come a long way since Spinoza. Today, multifocal lenses are commonplace. The *passage of nature* or *ground Gestalt of time* functions as a multifocal lens that allows us to identify nature in all that we see. It will enlighten our awareness if only we take the trouble to look through it.

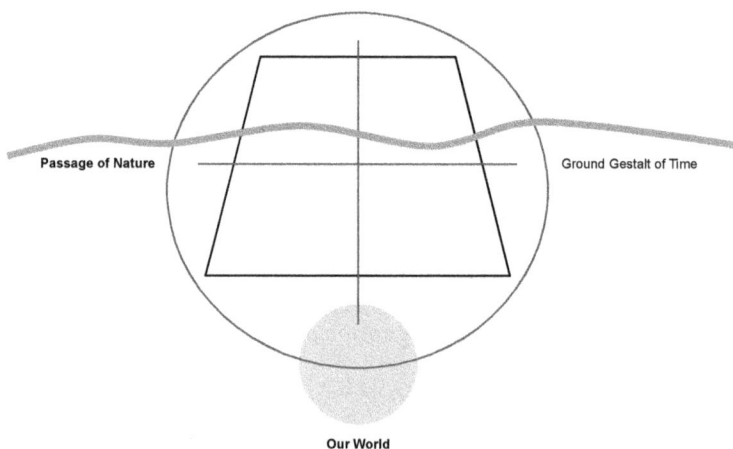

Figure 19 A new lens to identify nature in all that we see.

Notes

---

[1] Fabio Costa, Quantum causal relations: A causes B causes A, EurekAlert! (2 October 2012). http://www.eurekalert.org/pub_releases/2012-10/uov-qcr100212.php

---

[2] Marc van der Erve, *A New Dimension of Time*, Antwerp (2008).

[3] The expression "only together they arise" stems from "dependent arising" or "pratityasamupada" (in Sanskrit): http://en.wikipedia.org/wiki/Paticca-samuppada See also: Joanna Macy, *Mutual Causality in Buddhism and General Systems Theory*, State University of New York, New York, (1991).

[4] Vanessa Thorpe, *Richard Dawkins in furious row with EO Wilson over theory of evolution*, The Observer, (Sunday 24 June 2012).

[5] Marc van der Erve, *A New Leadership Ethos – The Ability to Predict*, Antwerp (2008). I referred to: Richard Dawkins, *Climbing Mount Improbable*, W.W. Norton & Company, New York (1996) and Richard Dawkins, *The Extended Phenotype*, Oxford University Press (1999).

[6] Marc van der Erve, *To Be Or To Become,* Antwerp (2011): *"The 73-year-old Swedish scientist, Lars Olov Bygren, studied the effect of feast and famine winters in very cold, isolated and sparsely populated areas that conveniently served as reaction tubes. Bygren's research revealed that the offspring of kids, who went from normal eating to overeating in rare warm and, thus, overabundant winters, produced offspring that died six years earlier on average. The research of Bygren demonstrated the link between temperature inequality, behavior patterns and the health of offspring, which evidently depends on DNA. However, assuming that the DNA-code remained unchanged, which it did, the question was how this could have happened? The findings of Bygren led to the suspicion that the temperature-inspired change of eating behaviors had either switched on or off certain DNA-code segments. It turned out that this is common practice inside the human body. Because all cell types contain the same DNA, local inequalities either silence or activate elements of DNA to help cells differentiate and grow into skin cells, organ cells, and so on. Just outside the DNA molecule, chemicals above the ladder of sequences (that is, the ladder seen on its side) function as "flags" that turn genes on or off. These chemicals are called the "epigenome", the prefix, epi, meaning above. The science that emerged from these findings is fittingly named "epigenetics." This is another example of how the behavioral view of our world is gradually being embraced."* I referred to: Lars Olov Bygren, *Longevity Determined*, Acta Biotheoretica, Volume 49, Number 1, (2001).

[7] Stephen Jay Gould, *Bully for Brontosaurus*, W. W. Norton, New York (1991).

[8] Simon Morris Conway, *Life's Solution*, Cambridge University Press (2003).

[9] (Source: IBM) *IBM Applies Self-Assembling Nanotechnology To Conventional Chip Manufacturing*, Nanowerk.com (3 May 2007). http://www.nanowerk.com/news/newsid=1883.php

[10] The 18-year old researcher, Hunter Browning, created a novel hydrogen cell that is efficient enough to power a cooking grill: Joe Robertson, *Blue Valley senior is on the verge of a quantum leap in science*, Kansas City Star (5 June 2011).

http://joco913.com/news/blue-valley-senior-is-on-the-verge-of-a-quantum-leap-for-science/

[11] Reversing time would violate the Second Law. The entropy generated the first time around cannot be reclaimed. It is simply lost.

[12] Arthur Koestler, *The Ghost in the Machine*, Arkana Books, London (1989), p. 45.

[13] James Kelleher, *In crisis, GE finds its deep bench not so magical*, Reuters (March 6, 2009). http://www.reuters.com/article/2009/03/06/us-ge-management-idUSTRE5253TC20090306

[14] I identified this for the first time in: Marc van der Erve, *Temporal Leadership*, European Business Review, Vol. 16, No. 6 (2004).

[15] Current-day examples are North Korea and Zimbabwe.

[16] This doesn't mean that management teams should consist of people all with the same leadership comfort zone (something that Welch managed to do). In view of future stages, management teams can be made up of people with different comfort zones. Only on average, the balance of comfort zones should tip to the comfort zone needed in the target stage.

[17] My extensive study of leadership comfort zones has shown that the decisions of leaders are dominated by one stage of organizational development.

[18] In the end, the idea to create stage-specific consulting practices to improve their effectiveness was deemed futile by a top-of-the-bill consulting-company partner who before resonated with the idea. In 2010, when I discussed similar ideas with the CEO of a prominent US business school, more or less the same happened. These executives failed to convince their hinterland.

[19] I hope to contribute to the establishment of such Internet-based services.

[20] An analysis of the longer-term share price evolution of companies, such as GM, GE, and Apple, confirmed the share price trend in each of the four stages of behavior-pattern development. Of course, the share-price trend is an effect (rather than cause) of the stages of behavior-pattern emergence. The four stages of behavior-pattern emergence also explain the "Elliott wave".

[21] Henri Bergson, *Time and Free Will*, Dover, London (2001).

[22] The overwhelming evidence that the perception of our world as material phenomenon is a reflection of a non-material behavioral phenomenon appears now to be supported by what is referred to as the physics of the "unparticle". The signature of our world as behavioral phenomenon hinges entirely on thermodynamics. Interestingly, "unparticle physics" entirely hinges on a thermodynamic signature too. Howard Georgi first related "missing energy

distributions" to the "unparticle" in: *Unparticle Physics*, arXhiv (20 May 2007): http://www.people.fas.harvard.edu/~hgeorgi/utalkc.pdf

[23] Benedict de Spinoza (Translation from Latin to English by R.H.M. Elwes, 1915), *Ethica.English* (*Ethica Ordine Geometrico Demonstrata*), Amsterdam (1677). The name given to Jewish-born Spinoza was "Baruch Spinoza". Expelled from the Jewish community in Amsterdam, Spinoza adopted the Christian name "Benedict de Spinoza" later in life. Spinoza died at the age of 44.

[24] Wikipedia article on Spinoza: http://en.wikipedia.org/wiki/Baruch_Spinoza

[25] As Stuart Hampshire notes, "*he [Spinoza] could not accept the simple-minded materialist conception*": Stuart Hampshire, *Spinoza and Spinozism*, Oxford University Press, New York, (2005).

[26] The quotes included are from Spinoza's book *Ethica*. Benedict de Spinoza (Translation from Latin to English by R.H.M. Elwes, 1915), *Ethica.English* (*Ethica Ordine Geometrico Demonstrata*), Amsterdam (1677).

[27] Stuart Hampshire notes about the view of Spinoza: "*reality is a single, self-generating system, natura naturans.*" Stuart Hampshire, *Spinoza and Spinozism*, Oxford University Press, New York, (2005).

[28] Spinoza emerged from his texts as a modern-thinking man locked into an era of the past.

Epilogue

In his book *The End of Time: The Next Revolution in Physics*, Barbour eagerly referred to a statement by his countryman and peer, Stephen Hawking. Hawking had apparently said that the ultimate "theory of everything" would appear within the next two decades or so. Barbour added to this: [1]

> In the next revolution in physics, [ordinary] time will cease to have a role in the foundation of physics.

Barbour was *so* right! *Ordinary time* or the *figure Gestalt of time* may indeed lose its role in the foundation of physics. Yet, the *passage of nature* or the *ground Gestalt of time* will replace it. Barbour seemed to confirm this indirectly, when he wrote,

> One day, the theory of evolution will be subsumed in a greater scheme, just as Newtonian mechanics was subsumed in Relativity.

Trends 1 to 5 in Part 4 suggest that this "greater scheme" involves not just the *passage of nature* or *ground Gestalt of time* but also a wholly new view of reality. Of course, Kuhn warned that such daring propositions are not easily accepted. Quoting the German physicist, Max Planck, he noted,[2]

A new scientific truth does not necessarily triumph by convincing its opponents and making them see the light but rather because its opponents eventually die and a new generation grows.[3]

I depicted a "new scientific truth", which a collective of scientists and thinkers, headed by Baruch Spinoza, identified when stumbling over the scientific and philosophical anomalies of their time. I did so by connecting their ideas. So, what are the chances that the *behavioral worldview* will be adopted? Kuhn offered criteria that, he admitted, should be used as indications only. To be accepted, he said, it should be *empirically adequate*, *consistent across theories*, *broad in scope*, *simple*, and *fruitful*. I hope to have shown in this book that the *behavioral worldview* meets these criteria. One might also ask whether someone from outside the scientific world, like me, is a credible source for such a "new scientific truth". In the following excerpt, Kuhn seemed to suggest that my endeavors might not have been in vain.

Men who will identify a paradigm shift typically are little committed by prior practice to the traditional rules of normal science and are likely to see that those rules no longer define a playable game and to conceive another set that can replace them.

Lastly, early in my career, I had the privilege of working for *Jack Schweizer*, senior vice president of *Digital Equipment Corporation* in Europe. An exceptional leader, Jack, much like Alexander the Great, would not waste his time unraveling a Gordian knot.[4] With brute force, he would wield a virtual sword during many a meeting to cut through confusing complexities and lay bare the core of the matter. I remember our discussions, in which he urged me to evaluate how a target audience might perceive the message presented because, as he reminded me often, "perception is reality". I have learned three crucial things from Jack. First, the common perception of our world is like a Gordian knot. It involves a history of convoluted assumptions that is nearly impossible to unravel. Second, to lay bare the core of the matter again, it is sometimes useful to wield a virtual sword to cut through this knot. Third, our perception is the place to start because, in the end, it determines what we see.[5]

Notes

---

[1] Julian Barbour, *The End of Time: The Next Revolution in Physics*, Oxford University Press (1999).

[2] Thomas Kuhn, *The Structure of Scientific Revolutions*, University of Chicago (Third edition, 1996).

[3] This is why I also targeted undergraduate and post-graduate students.

[4] Alexander the Great did probably not cut through a Gordian knot but instead pulled the knot out of its pole pin. This exposed the two ends of the rope and allowed him to untie the knot: E. Fredricksmeyer, *Alexander, Midas, and the Oracle at Gordium*, Classical Philology, Vol. 56, No. 3 (July, 1961), pp. 160-168.

[5] Jack died in October 2012, the month in which I finished writing this book.

Acknowledgements

I am grateful to nature for allowing me to get *this* close. By now, I know that this may not have been my choice but my destiny (Part 4, Trend 19). I am also grateful to Jack Hydes and Hubert Kals for their most constructive comments. Lastly, I am indebted to my friends, my kids in law, my kids and, especially, my wife, Karen, for bearing with me when pursuing my research.

Marc van der Erve

Bibliography

IBM Applies Self-Assembling Nanotechnology To Conventional Chip Manufacturing, Nanowerk.com (3 May 2007).

How rocks evolve, The Economist (13 November 2008).

Lifetime of Average S&P Company, Business Innovation Insider (2005), No. 11.

Schumpeter | *Built to last*, The Economist (26 November 2011).

The Higgs Boson: Fantasy turned reality. The Economist (14 December 2011).

The Skeleton of Water, The Economist (14 November, 2009).

Aristotle, *On Physics*: http://classics.mit.edu/Aristotle/physics.mb.txt

Ross Ashby, *Principles of the self-organizing system*, in: *Principles of Self-Organization: Transactions of the University of Illinois Symposium*, Heinz Von Foerster et al, Pergamon Press, London, (1962).

Jan Assmann, *Moses The Egyptian*, Harvard University Press (1998).

Per Bak, *How Nature Works: The Science of Self-Organized Criticality*, Copernicus, New York (1996).

Julian Barbour, *The Development of A New Dimension of Time Themes in the Twentieth Century*, in: Jeremy Butterfield, *The Arguments of Time*, Oxford University Press (first edition, 1999, paperback 2006).

The Next Scientific Revolution

Julian Barbour, *The End of Time: The Next Revolution in Physics*, Oxford University Press (1999).

John Barrett, Harald Garcke, Robert Nürnberg, *Numerical computations of facetted pattern formation in snow crystal growth*, arXiv (6 February 2012).

Gregory Bateson, *Mind and Nature - A Necessary Unity*, Bantam Books, London (1979).

Gregory Bateson, *Steps to an Ecology of Mind*, University of Chicago Press (1972)

Adrian Bejan, *Advanced Engineering Thermodynamics*, Wiley, New York (1997).

Adrian Bejan, *Shape and Structure, from Engineering to Nature*, Cambridge University Press (2000).

Roberto Benzi, *Stochastic Resonance: From Climate to Biology*, Nonlinear Processes in Geophysics, 17, (2010), pp. 431–441.

Henri Bergson, *Time and Free Will: An essay on the immediate data of consciousness*, Dover edition, Mineola, New York (2001).

Paul Bloom, *The Original Colonists*, The New York Times (11 May 2012).

Ludwig Boltzmann, *The second law of thermodynamics*, Populare Schriften, Essay 3 (29 May 1886).

Frank Borg, *What is osmosis? Explanation and understanding of a physical phenomenon*, (2003). http://arxiv.org/abs/physics/0305011

Jerome Burner, Leo Postman, *On the Perception of Incongruity: A Paradigm*, Journal of Personality, 18, pp. 206-223 (1949).

Jeremy Butterfield, Chris Isham, *On the Emergence of Time in Quantum Gravity*, in: Jeremy Butterfield, *The Arguments of Time*, Oxford University Press (first edition, 1999, paperback 2006).

Lars Olov Bygren, *Longevity Determined*, Acta Biotheoretica, Volume 49, Number 1, (2001).

Scott Camazine, Jean-Louis Deneubourg, Nigel Franks, James Sneyd, Guy Theraulaz, Eric Bonabeau, *Self-organization in Biological Systems*, Princeton University Press (2001).

James Collins, Jerry Porras, *Built To Last – Successful habits of visionary companies*, HarperCollins Publishers, New York (1994).

Simon Morris Conway, *Life's Solution*, Cambridge University Press (2003).

Fabio Costa, Quantum causal relations: A causes B causes A, EurekAlert! (2 October 2012).

Ron Cowen, *Snowflakes Re-created Using Physics*, Scientific American (18 March 2012).

Bibliography

Marie Jean Antoine Nicolas Caritat de Condorcet (Marquis de Condorcet), *Esquisse d'un Tableau Historique des Progrès de l'Esprit Humain*, Paris (1794).

Charles Darwin, *On the Origin of Species - by Means of Natural Selection or the Preservation of Favored Races in the Struggle for Life*, John Murray, London (1859).

Richard Dawkins, *Climbing Mount Improbable*, W.W. Norton & Company, New York (1996)

Richard Dawkins, *The Extended Phenotype*, Oxford University Press (1999).

Arie de Geus, *The Living Company*, Harvard Business School Press, Boston (1997).

Benedict de Spinoza (Translation from Latin to English by R.H.M. Elwes, 1915), *Ethica.English (Ethica Ordine Geometrico Demonstrata)*, Amsterdam (1677).

Ellen De Rooij, *A brief desk research study of the average life expectancy of companies in a number of countries*, Stratix Consulting Group, Amsterdam (August 1996).

Peter Drucker, *Innovation and Entrepreneurship*, Pan Books, London (1985).

Albert Einstein, *Relativity: The Special and General Theory* (Revised edition, 1924).

Frank Eltman, *Evolution skepticism will soon be history*, Independent Online (29May 2012).

E. Fredricksmeyer, *Alexander, Midas, and the Oracle at Gordium*, Classical Philology, Vol. 56, No. 3 (July, 1961), pp. 160-168.

Luca Gammaitoni et al, *Stochastic Resonance*, Reviews of Modern Physics (January 1998), Vol. 70, No. 1.

Owen Gingerich, *The Book Nobody Read*, Walker (2004).

Howard Georgi, *Unparticle Physics*, arXhiv (20 May 2007), http://www.people.fas.harvard.edu/~hgeorgi/utalkc.pdf

William Godwin, *Enquiry concerning Political Justice and Its Influence on Morals and Happiness*, Robinson, 3rd Edition (1789)

Stephen Jay Gould, *Bully for Brontosaurus*, W. W. Norton, New York (1991).

Brian Gongol, *Age of the World's Largest Companies* (2005).

Brian Greene, *The Elegant Universe*, W. W. Norton & Company, New York (2003).

Hermann Haken, Synergetics - Introduction and Advanced Topics, Springer, Berlin (2004).

George Haller, *Chaos Near Resonance*, Springer, New York (1999).

George Haller, *Lagrangian coherent structures from approximate velocity data*, Physics of Fluids, Vol. 14, Number 6 (June 2002).

Stuart Hampshire, *Spinoza and Spinozism*, Oxford University Press, New York, (2005).

Sam Harris, *The Moral Landscape*, Free Press, New York, 2010.

Robert Hazen et al, *Mineral evolution*, American Mineralogist 93, pp. 1693-1720 (2008).

Gary Herstein, *Alfred North Whitehead (1861-1947)*, IEP (8 May 2007).

Tim Hornyak, *British atomic clock is world's most accurate*, CNet News (27 August, 2011).

William Huitt at al., *Piaget's theory of cognitive development*, Educational Psychology Interactive, Valdosta State University (2003).

Walter Isaacson, Steve Jobs, Simon & Schuster, New York (2011).

Chris Isham, Konstantina Savvidou, *Time and modern physics*, pp. 6-26 in: Katinka Ridderbos (editor), *Time*, Cambridge University Press (2002).

Chris Isham, *Topological and global aspects of quantum theory*, in: B. S. DeWitt, R. Stora (editors), *Relativity, Groups and Topology II*, Proceedings of the 40th Summer School of Theoretical Physics, NATO Advanced Study Institute, Les Houches, France (27 June- 4 August 1983), pp. 1059-1290, Amsterdam, New York (1984).

Kunihiko Kaneko, Ichiro Tsuda, *Complex Systems: Chaos and Beyond*, Springer-Verlag, Berlin (2000).

Kunihiko Kaneko, Ichiro Tsuda, *Chaotic Itinerancy*, Chaos (in: Focus Issue on Chaotic Itinerancy), 13(3)(2003), pp. 926-936.

Kunihiko Kaneko, *Dominance of Milnor attractors in globally coupled dynamical systems with more than 7 degrees of freedom*, Physical Review, E 66, 055201(R) (2002).

Stuart Kauffman, Lee Smolin, *A Problem with the Argument of Time*: http://www.edge.org/3rd_culture/smolin/smolin_p4.html

Stuart Kauffman, *The Origins of Order: Self-Organization and Selection in Evolution*, Oxford University Press, New York (1993).

James Kelleher, *In crisis, GE finds its deep bench not so magical*, Reuters (March 6, 2009).

Hitoshi Kitada, Lancelot Fletcher, *Comments on the Problem of Time*, (22 August 1997).

Michael Klesius, *The Mystery of Snowflakes*, National Geographic (January 2007).

Bibliography

Arthur Koestler, *The Act of Creation*, Arkana Penguin Books, London (1964, this edition: 1989).

Arthur Koestler, *The Ghost in the Machine*, Arkana Books, London (1967, this edition: 1989).

Vladimir Koutvitsky, Eugene Maslov, *Instability of coherent states of a real scalar field* (12 October 2005).

Vladimir Koutvitsky, Eugene Maslov, *Gravipulsons*, American Physical Society, Phys. Rev. D, Vol. 83, No. 12 (17 June 2011).

Thomas Kuhn, *The Structure of Scientific Revolutions*, University of Chicago (Third edition, 1996).

Bernard Lietaer et al, *Money and Sustainability – The Missing Link*, Triarchy Press, Axminster (2012).

Kenneth Libbrecht, *Observations of an Edge-enhancing Instability in Snow Crystal Growth near -15 C*, arXiv (11 November 2011.

Kenneth Libbrecht, *The Physics of Snow Crystals*, Institute of Physics Publishing, Reports on Progress in Physics, 68 (2005).

Nikolay Lossky, *The Intuitive Basis of Knowledge* (original in 1903, translation in 1919).

Joanna Macy, *Mutual Causality in Buddhism and General Systems Theory*, State University of New York, New York, (1991).

Thomas Robert Malthus, *An Essay on the Principle of Population, as it affects The Future of Society with remarks on the speculations of Mr. Godwin, M. de Condorcet and other writers*, J. Johnson, London (1789).

Benoît Mandelbrot, *The fractal geometry of nature*, W.H. Freeman, New York (1983).

Cesare Marchetti, *Kondratiev Revisited – After One Kondratiev Cycle*, International Institute for Applied Systems Analysis, Laxenburg (1988).

Lynn Margulis, Dorion Sagan, *What is Life?* University of California Press, Berkley (1995, 2000).

Lynn Margulis, Dorion Sagan, Marvelous microbes, Resurgence 206 (2001).

Lynn Margulis (previously Sagan), *On the origin of mitosing cells*, Journal of Theoretical Biology, Vol. 14, Issue 3 (March 1967)

Eugene Maslov, *Parametric Resonance As Possible Cause Of Spontaneous Transition From Meta-stable States*, Annales de la Fondation Louis de Broglie, Vol. 26 Spécial, (2001).

Humberto Maturana, Francisco Varela, *Autopoiesis: The Organization of the Living*, D. Reidel Publishing Company, Dordrecht (1972).

Humberto Maturana, Francisco Varela, *The Tree of Knowledge – The Biological Roots of Human Understanding*, Shambhala Publications, Boston (1992).

J. H. McCulloch, *The Austrian Theory of Marginal Use and of Ordinal Marginal Utility*, Journal of Economics, Springer Verlag (1977), Vol. 37, No. 3-4, pp. 249-280).

Theodore Modis, *Predictions - Society's Telltale Signature Reveals the Past and Forecasts the Future*, Simon & Schuster, New York (1992).

Clara Moskowitz, *Quantum mechanics on steroids: Even the largest molecules behave like waves*, MNN (27 March 2012).

Rosabeth Moss-Kanter, *The Change Masters*, Unwin Hyman, London (1989).

Robert Mulligan, *An Empirical Examination of Austrian Business Cycle Theory*, The Quarterly Journal of Austrian Economics (Summer 2006), Vol. 9, No. 2, p. 69-93.

Lars Onsager, *Reciprocal Relations in Irreversible Processes* I., Physical Review No. 37, pp. 405-426 (1931).

Marcel Pawlowski, Jan Pflamm-Altenburg, Pavel Kroupa, *The VPOS: a vast polar structure of satellite galaxies, globular clusters and streams around the Milky Way*, Monthly Notices of the Royal Astronomical Society, 000, 1-21 (2012).

Roger Penrose, *Cycles of Time: An Extraordinary New View of the Universe*, Alfred A. Knopf, New York (2011).

Roger Penrose, Fedja Hadrovich, *Twistor Theory*, http://users.ox.ac.uk/~tweb/00006/index.shtml

Tom Peters, *Thriving on Chaos: Handbook for a Management Revolution*, Alfred A. Knopf, New York (1987).

Jean Piaget, *Genetic Epistemology*, Columbia University Press (1968).

Ilya Prigogine, *Time, Structure, and Fluctuations*, Nobel Lecture (1977).

Ilya Prigogine, Isabelle Stengers, *The End of Certainty*, The Free Press, New York (1996).

Gilbert Probst, Hans Ulrich, *Self-Organization and Management of Social Systems*, Springer Verlag, Heidelberg (1984).

Conrad Quilty-Harper, *Steve Jobs at Apple: a relentless rise in graphs and charts*, The Telegraph (6 October, 2011).

Matt Ridley, *The Origin of Virtue*, Viking (1996).

Joe Robertson, *Blue Valley senior is on the verge of a quantum leap in science*, Kansas City Star (5 June 2011).

Bibliography

Carlo Rovelli, *Forget time*, Essay written for the FQXi contest on the Nature of Time (August 24, 2008).

Carlo Rovelli, Loop *Quantum Gravity*, Living Rev. Relativity, 11 (2008), 5.

Carlo Rovelli, *The First Scientist: Anaximander and His Legacy*, Westholme, Yardley (2011).

Erwin Schrödinger, *What is Life? – The Physical Aspect of the Living Cell*, (1944).

Joseph Alois Schumpeter, *Capitalism, Socialism and Democracy* (1942).

Adam Smith, *An Inquiry into the Nature and Causes of the Wealth of Nations*, London (1776).

Lee Smolin, *Three Roads to Quantum Gravity*, Basic Books, New York (2001).

Lee Smolin, *The present moment in quantum cosmology: Challenges to the arguments for the elimination of time* (30 August, 2000).

Lee Smolin, *The Trouble With Physics: The Rise of String Theory, The Fall of a Science, and What Comes Next*, Houghton Mifflin Company, Boston (2007).

Steven Strogatz, *Sync: The Emerging Science of Spontaneous Order*, Hyperion, New York (2003).

Rod Swenson, Autocatakinetics, Yes – Autopoiesis, No: Steps Toward A Unified Theory of Evolutionary Ordering, International Journal of General Systems, Vol. 21 (1992).

Rod Swenson, T*he Fourth Law of Thermodynamics or the Law of Maximum Entropy Production*, Chemistry, Vol. 18, Issue 5 (2009).

Wenbo Tang, Pak Wai Chan, George Haller, *Lagrangian Coherent Structure Analysis of Terminal Winds Detected by Lidar. Part II: Structure Evolution and Comparison with Flight Data,* Journal of Applied Meteorology & Climatology, Vol. 50, Issue 10, pp. 2167-2183 (October 2011).

Vanessa Thorpe, *Richard Dawkins in furious row with EO Wilson over theory of evolution*, The Observer, (Sunday 24 June 2012).

Ichiro Tsuda, *Hypotheses on the functional roles of chaotic transitory dynamics*, Chaos, No. 19, 015113-1 – 015113-10 (2009).

Ichiro Tsuda, *Toward an interpretation of dynamic neural activity in terms of chaotic dynamical systems*, Behavioral and Brain Sciences, 24(5) (2001).

Brian Vastag, Joel Achenbach, *Scientists discover new subatomic particle at the center of everything*, The Washington Post (4 July 2012).

Marc van der Erve, *A New Dimension of Time*, Antwerp (2008).

Marc van der Erve, *A New Leadership Ethos – The Ability to Predict*, Antwerp (2008).

Marc van der Erve, *Evolution Management – Winning in Tomorrow's Marketplace*, Butterworth-Heinemann, Oxford (1994).

Marc van der Erve, *Dynamisch Ondernemen – Strategieën voor de ontwikkeling van een flexibele organisatie*, Sijthoff, Amsterdam (1986).

Marc van der Erve, *Resonant Corporations*, McGraw-Hill, New York (1998).

Marc van der Erve, *Temporal Leadership*, European Business Review, Vol. 16, No. 6 (2004).

Marc van der Erve, *The Mathematical Model and Simulation of a Nonlinear Real-World System*, Werktuigkundig Laboratorium voor Meet- en Regeltechniek, TU Delft (1974).

Marc van der Erve, *The Power of Tomorrow's Management*, Heinemann Professional Publishing, Oxford (1989).

Marc van der Erve, *To Be Or To Become*, Antwerp (2011).

Francisco Varela, *Two Principles of Self-Organization*, in: Gilbert Probst, Hans Ulrich, *Self-Organization and Management of Social Systems*, Springer Verlag, Berlin (1984).

Pierre-François Verhulst, *Notice sur la loi que la population poursuit dans son accroissement*, University of Ghent (1838).

Erik Verlinde, *On the Origin of Gravity and the Laws of Newton* (6 January 2010). http://arxiv.org/pdf/1001.0785

Tamás Vicsek, *Fractal Growth Phenomena*, World Scientific Publishing, Singapore, London (1992).

Alfred Russel Wallace, *My Life*, Chapman & Hall (1905).

Alfred Russell Wallace, *On the Law which has Regulated the Introduction of New Species*, Annals and Magazine of Natural History (1855).

Alfred Russel Wallace, *On the Tendency of Varieties to Depart Indefinitely From the Original Type* (1855).

Susan Watts, *CERN scientist expects "first glimpse" of Higgs boson*, BBC News (7 December 2011).

Elizabeth Weise, *Scientist wants to make snowflake formation crystal clear*, USA Today (18 December 2011).

Richard Wendt, *Simplified Transport Theory for Electrolyte Solutions*, Journal of Chemical Education, Vol. 51, p. 646 (1974).

Alfred North Whitehead, *Process and Nature*, The Free Press, New York (1978).

Alfred North Whitehead, *The Concept of Nature*, Prometheus Books, Amherst (2004).

Bibliography

Edward Wilson, *The Social Conquest of Earth*, Liveright Publishing Corporation, New York (2012).

Peter Woit, *Not Even Wrong: The Failure of String Theory*, Basic Books, New York (2006).

Stephen Wolfram, *A New Kind of Science*, Wolfram Media (2002).

Colin Woodard, *Book Review: "The Social Conquest of Earth," by Edward O. Wilson*, The Washington Post (13 April 2012).

Index

Bergson
  Henri, 72, 73, 135, 141, 160
Big Bang, 90
Big Bounce, 90
Borg
  Frank, 95
Bruner
  Jerome, 20
Buddha
  Gautama, 147
builder-type leader, 13

C

catalysts of recognition, 21
causal connection, 87
cause, 36
chaos, 66
Clausius
  Rudolf, 60
closure, 58
cognitive development, 29
coherence, 58
Collins
  Jim or James, 40
commensurable, 28
concrete slab of nature, 75
congruent simultaneity, 124, 138
connectio-dynamic, 130
connectio-dynamics, 82
constructal law, 24, 63
convection cells, 58
convergent evolution, 150
Conway Morris
  Simon, 151
Copernican Revolution, 19
Copernicus
  Nicolaus, 18
corpuscle, 16
creation of forms, 73
creative destruction, 41

Cro-Magnon, 134

D

dark matter, 97
Darwin
  Charles, 52
Dawkins
  Richard, 53, 149
de Condorcet
  Nicolas, 47
de Geus
  Arie, 70
Dekker
  Wisse, 37
Delft University of Technology, 37
Descartes
  René, 16, 154
Digital Equipment, 36
diminishing returns, 50
diversity, 148
division of labor, 127
dogma of materialism, 135
dualism, 154
duration, 73, 137, 141

E

Einstein
  Albert, 16
electron Volts, 15
endosymbiosis, 62
energy inequality, 57
energy niche, 25
entropy, 60, 81, 85
equality, 154
equilibrium ocean, 89, 96, 119
events, 76

F

G

H

I

Z